FIREFLY

GUIDE TO THE OCEANS

DR. JOHN PERNETTA

FIREFLY BOOKS

A FIREFLY BOOK

Published by Firefly Books Ltd. 2004

Second printing, 2006

Publisher Cataloging-in-Publication Data (U.S.)
Pernetta, John.
Guide to the oceans / John Pernetta. – 1st e
[240] p. : col. ill., photos. , maps ; cm.
Includes index.
Summary: Illustrated guide to the features of ocea and seas, including an encyclopedia of marine life.
ISBN-13: 978-1-55297-942-6 (pbk.)
ISBN-10: 1-55297-942-3 (pbk.)
1. Ocean. 2. Marine life. 3. Marine biology. I. Titl
551.46 22 GC21.P47 2004

National Library of Canada Cataloguing in Publication Data
Pernetta, John
Guide to the oceans / John Pernetta.
Includes index.
ISBN-13: 978-1-55297-942-6
ISBN-10: 1-55297-942-3
1. Ocean. 2. Ocean--Maps. 3. Marine biology. I. Ti
GC21.P47 2004 551.46 C2003-907132-4

Published in the United States by
Firefly Books (U.S.) Inc.
P.O. Box 1338, Ellicott Station
Buffalo, New York 14205

Published in Canada by
Firefly Books Ltd.
66 Leek Crescent
Richmond Hill, Ontario L4B 1H1

Printed in China

COMMISSIONING EDITOR *Christian Humphries*
EDITOR *Joanna Potts*
EXECUTIVE ART EDITOR *Mike Brown*
DESIGNER *Chris Bell*
PRODUCTION *Sally Banner*

FRONT COVER : top left: hatchet fish, © Philip's; top right: Caribbean Sea map, © Philip's; center: globe with oceanic trenches, © Philip's; bottom left: yachts, Gary John Norman/ Stone/*Getty Images;* bottom right: scuba diver, Peter Scoone *Science Photo Library*
BACK COVER : bottom left: puffin, © Philip's; bottom right: co formation, © Philip's
THIS PAGE (from top to bottom) : wave, Royalty free/*Corbis;* coral reef, Florida Keys National Marine Sanctuary/*NOAA;* windsurfers, Royalty free/*Getty Images;* nets at Cochin, southern India, Royalty free/*Corbis*

Contents

The Oceans

GENESIS OF THE OCEANS

In Greek mythology, the great river or sea that surrounded the flat disk of the Earth, was personified by Oceanus, the god of the primeval waters. Now, of course, we know that the world is spherical and that slightly more than 70 percent of its surface is covered by the oceans.

From space, the blue of the Earth reflects the vast expanse of water covering its surface. From a point in outer space above the center of the Pacific Ocean, the entire planet seems to be composed of water, since no land is visible. The dominance of the ocean has earned Earth the title of the 'Blue Planet'.

The oceans are by far the largest store of water on the planet. In comparison, the freshwater contained in icecaps, rivers, lakes and groundwater is much smaller, while the clouds and water vapor in the atmosphere make up only a minute fraction of the total. Water leaves the oceans through evaporation and enters via rainfall, rivers and groundwater seepage. The total volume of water involved in this continuous cycle is approximately 108,000 cubic miles (450,000 cubic kilometers).

Surprisingly, the Earth's crust also contains significant amounts of water. The water either permeated into the underlying rocks or was incorporated into sediments during their formation. In areas where volcanic activity occurs, the

molten rock (magma) absorbs water as it rises toward the surface of the crust where it is released as superheated steam during volcanic eruptions. It seems that the water in the crust is critical to the formation of magma, since it lowers the melting point of the rock at depth.

▲ ***The oceans*** *are the largest store of water on the planet. Four thousand million years ago, the atmosphere was dominated by water vapor. As the Earth cooled, condensing water vapor fell as rain, forming rivers and filling the lower-lying areas to form the oceans.*

Ocean formation

Most of the water now lying in the oceans has been on Earth since very early in geological history, mainly in the form of water vapor in the atmosphere. As the Earth cooled, the condensing water vapor fell as rain, forming rivers and filling the lower-lying areas to form the oceans. Although we cannot be certain as to when this process took place, geological evidence suggests that surface waters existed 3,800 million years ago. The evidence takes the form of the oldest-known sedimentary (water-formed) rock, discovered in Greenland. While evidence of simple organisms has been found in types of rock 3400 to 2000 million years old, the earliest traces of multicelled animals are a mere 600 million years old.

It is in the oceans that life began. Initially it took the form of simple organic molecules, followed later by self-replicating, large organic molecules – the nucleic acids. These would eventually provide the mechanism of reproduction, enabling cells to divide and give rise to new cells like themselves.

◄▼ ***These satellite** images of the Earth show its great expanse of ocean, the reason for it being called the 'Blue Planet'. The globe on the left shows the western hemisphere, with the Pacific Ocean to the west and the Atlantic Ocean to the east. The Pacific Ocean is the world's largest ocean. The globe below shows the eastern hemisphere with the Indian subcontinent at the center and the Indian Ocean below.*

For nearly 2000 million years, simple, single-celled microorganisms were the only form of life on planet Earth. Their existence depended on the presence of water, and it was their life processes that altered the composition of the atmosphere. At first, Earth's atmosphere was dominated by water vapor, methane and ammonia, but the mere existence of photosynthesising microorganisms led to increasing concentrations of oxygen in the atmosphere. This oxygen led in turn to the formation of the ozone shield which now protects life on Earth from the harmful effects of the Sun's ultraviolet radiation.

The changing oceans

The ocean basins have not always had their present form. The movement of the continents and ocean-floor spreading have resulted in periods when the oceans were quite differently arranged. Some 225 million years ago, the land formed a single, large continental land mass called Pangaea, which was surrounded by a superocean, Panthalassa. When this land mass broke up some 180 million years ago to form two super continents (Laurasia to the north and Gondwanaland to the south), the Tethys Sea was formed between them. This pantropical sea was subsequently closed by the collision of what is now Africa with the Eurasian mainland. The present ocean basins are thus relatively young, being less than 80 million years old, and the oldest rocks under the ocean floor are a mere 200 million years old compared with the oldest igneous rocks on land, which are some 4200 million years old.

The ocean floor has a remarkable topography, with vast plains covered in a thick layer of sediment, and huge mountain ranges, the midoceanic ridges where upwelling (vertical water movement) from the mantle below pushes the ocean crust toward the surface. Some continents are bounded by wide and comparatively shallow continental shelves, while in areas where the ocean mantle descends beneath the continental margins, deep trenches such as the Peru-Chile trench are found. Where major rivers empty into the sea, their sediment is deposited near the mouths of the rivers to form huge underwater deltas, fans of sediment that spread out on the continental shelf, or cascade over the edge of the continental margin to the abyss below.

The ocean environment

The oceans are in constant motion. They are influenced by the rotation of the Earth, the gravitational pull of the Sun, the Moon and nearby planets, and by the movements of the atmosphere. This motion has important consequences, since the oceans circulate heat from the warmer tropics to the colder, higher latitudes, so influencing the climate of neighboring landmasses. Ocean circulation is not only an important component of the global climate system, but also results in quite different physical and chemical conditions for life in different parts of the ocean basins.

The environment of the ocean varies drastically. Conditions range from the warmer, sunlit surface waters to the dark and colder ocean deeps, and from the high-salinity ocean basins such as the Red Sea to the low-salinity conditions of the Bay of Bengal.

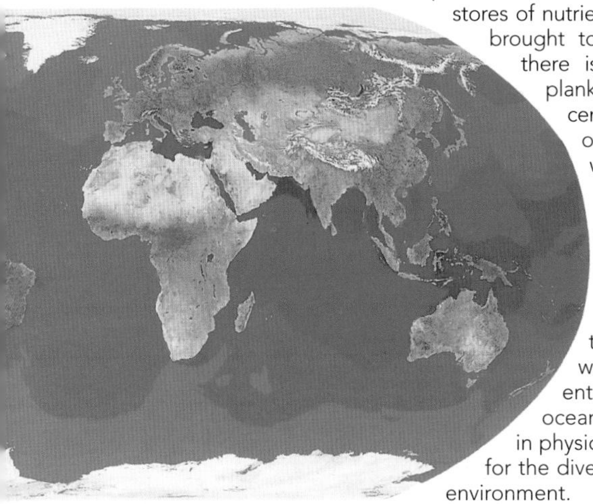

▼ ***This image of the Earth*** *shows the land in natural tones, the cooler ocean areas in light blue, and warm areas in darker tones.*

The surface waters of the ocean are generally low in nutrients, particularly in tropical areas, and this limits the rate of primary production by phytoplankton. Phytoplankton are single-celled plants that use sunlight to form more complex chemicals and which serve as the basis for marine life. The deep ocean, on the other hand, contains vast stores of nutrients, and where these are brought to the surface upwelling, there is a high production of plankton and fish. High concentrations of nutrients also occur in coastal regions where surface run-off carries nitrogen, phosphorus and organic material to the coastal waters.

This spatial variation in physical and chemical conditions influences the types of plant and animal which can survive in different sections of the world's ocean basins. It is this variation in physical features that accounts for the diversity of life in the marine environment.

▲ ***The size of the plates*** *is not constant. The African plate is bounded on three sides by midocean ridges that indicate the presence of constructive margins. The plate is constantly enlarging from these ridges, and there is no intervening destructive margin, therefore the ridges are moving apart as the plate enlarges.*

Planet Earth is made up of several concentric spheres of slightly different composition. The crust, or outer shell, consists of relatively light materials and is separated from the denser, underlying mantle by a region known as the Mohorovičić discontinuity. Beneath the continents, the Earth's crust is much thicker, 19–25 miles (30–40 kilometers), than beneath the ocean basins, where it is some 4 miles (6 kilometers) thick. In addition, the thickness of the ocean crust is essentially constant throughout the ocean basins. In contrast, the thickness of the continental crust varies considerably – the greatest depth being found beneath mountain ranges.

From mantle to core

The mantle extends to a depth of about 2900 kilometers (1800 miles) and is thought to consist of ferromagnesian silicate minerals such as olivene and pyroxene. The division between the upper section and lower, denser section of the mantle occurs at a depth of about 435 miles (700 kilometers). The upper mantle is in turn divided into the upper lithosphere, which extends down for between 30 and 155 miles (50 and 250 kilometers). Below this lies the asthenosphere, a partially molten zone some 125 miles (200 kilometers) thick. The asthenosphere defines the lower boundary of the relatively rigid outer lithosphere, which includes the crust. The plates of the lithosphere move over the semimolten asthenosphere, giving rise to the phenomenon of continental drift.

Below the asthenosphere lies the mesosphere, which is separated, at a depth of around 250 miles (400 kilometers) by a transition zone, from the lower mantle, which lies between 600 and 1200 miles (1000 and 2000 kilometers)

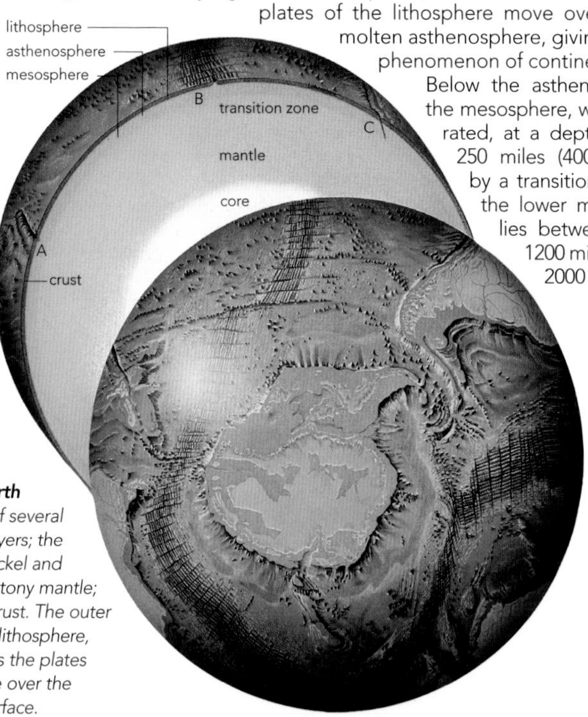

lithosphere
asthenosphere
mesosphere
B
transition zone
C
mantle
core
A
crust

► ***The Earth*** *consists of several distinct layers; the core of nickel and iron; the stony mantle; and the crust. The outer layer, the lithosphere, represents the plates that move over the Earth's surface.*

in depth. The lower boundary of the mantle is marked by the Gutenberg discontinuity, which separates the lower mantle from the underlying dense core. The core itself is divided into an inner and outer layer at a depth of between 3090 and 3170 miles (4980 and 5120 kilometers). The core most probably consists of an iron-nickel alloy, with a liquid outer core and an inner solid core with a radius of about 800 miles (1300 kilometers).

Seafloor spreading

Large-scale convection currents occur in the asthenosphere, and where these currents rise toward the Earth's surface, midocean ridges are formed. Where the currents flow toward the Earth's core, deep ocean trenches are created. The thin, oceanic crust forms at midocean ridges and is absorbed back into the mantle beneath the trenches. The youngest crust is, therefore, found near the center of the midoceans, and the oldest in the vicinity of the deep ocean trenches, which are generally found along the ocean margins. It is now thought that the oldest ocean crust is no more than 200 million years old.

New oceanic crust is formed on each side of the mid-oceanic ridges as the two plates move apart. Spreading rates vary in different ocean basins; for example, the movements away from the mid-Atlantic Ridge have been estimated at less than 0.4 inches (1 centimeter) per year, while the spreading along the East Pacific Rise is believed to be approximately 6.3 inches (16 centimeters) per year. As new crust is formed, it pushes the older crust further away from the ridge. The mid-oceanic ridges range from 0.6–2.5 miles (1 to 4 kilometers) in

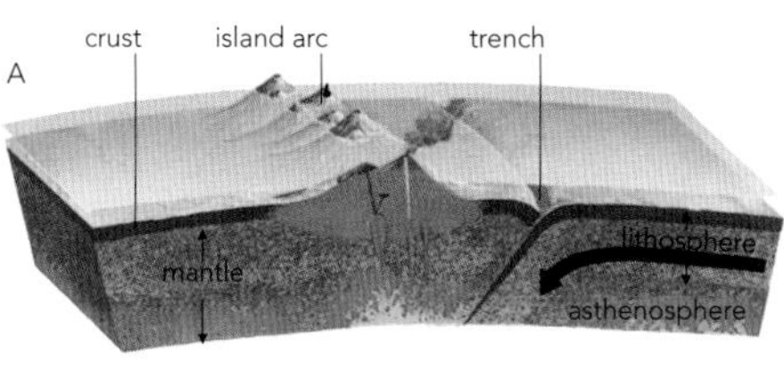

◀ ***The Tonga Trench*** *is a destructive plate margin, involving two oceanic plates. The margin is characterized by a deep trench and an arc of volcanic islands formed by melting within and above descending plates.*

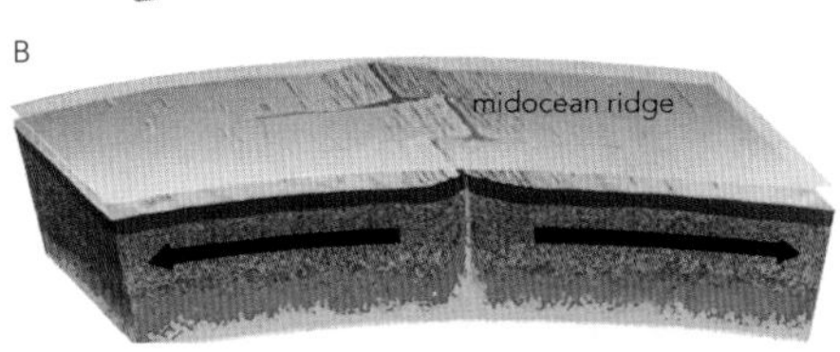

◀ ***The East Pacific Rise*** *is a constructive margin, where two oceanic plates are being continuously generated. Material is constantly rising to the surface, to be added to the edge of the plates as they move apart.*

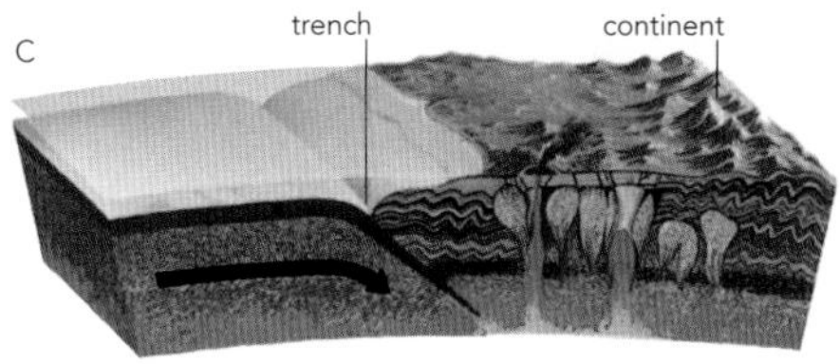

◀ ***The Peru-Chile Trench*** *is a destructive margin, where one oceanic and one continental plate meet. The oceanic plate forms a trench as it is carried down. The lighter continental material rises into a mountain range.*

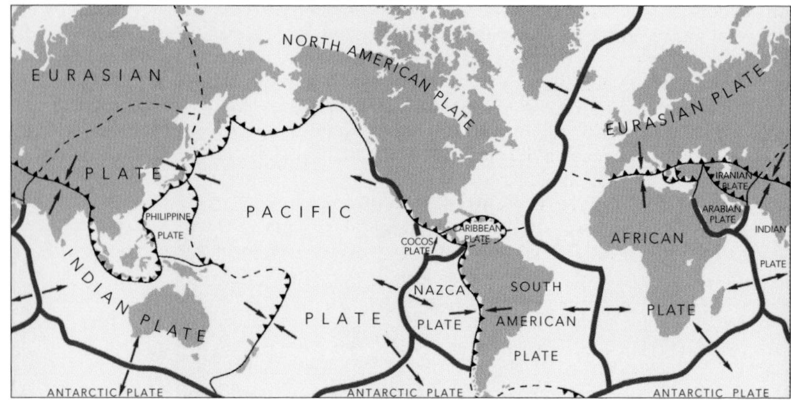

▲ ***Map of the Earth*** *depicting the position, the direction of movement, and type of plate boundaries.*

height, and along the center of the ridges lies the central rift where molten lava rises up from the mantle below. Lava also emerges from smaller vents along the sides of the midocean ridges, forming submarine volcanoes, which may break the ocean surface as volcanic islands. Chains of islands, with the older ones being the furthest from the ridge, may be formed as the two ocean plates move apart.

The midocean ridges are crossed by great transform faults which result in the ridges being offset, reflecting the irregularities of the edges of the plates. Along these transform faults, earthquakes are generated where newly formed parts of adjacent crustal plates slide past one another in opposite directions.

Hydrothermal vents, which are associated with areas of seafloor spreading, may produce metalliferous muds. These muds are formed from the precipitation of metals in solution when the hot molten materials from the underlying crust come into contact with the cooler ocean water. Hydrothermal vents often support a characteristic community of organisms,

◀ ***Eruption of Stromboli****, an island-volcano, the most northerly of the Lipari Islands, off the northeast coast of Sicily, southern Italy. Material ejected from Stromboli spills into the surrounding sea rather than falling back into the vent. This accounts for the fact that the island has been in a state of perpetual eruption for around 2500 years.*

which depend for their existence on bacteria that convert chemical energy into living matter. The bacteria in turn form the basis for the food chain of filter-feeding animals that comprise such communities.

The creation of new ocean crust along the midocean ridges is compensated for by its destruction in the area of the deep ocean trenches along the continental margins. In these areas the continental plate composed of less dense material, overrides the denser oceanic plate, which is forced downward and absorbed into the molten material at depths of around 435 miles (700 kilometers). Ocean trenches in these subduction zones may be as deep as 6.2 miles (10 kilometers), and the nearby continental margin is often characterized by intense volcanic activity such as in the Pacific 'Ring of Fire'.

Sedimentary material is added to the surface of the ocean crust as planktonic plants and animals die and their skeletons sink to the ocean floor. The rate of accumulation of these fine sediments is extremely slow, ranging from a few centimeters to as little as a fraction of a millimeter per thousand years. Despite such low rates of accumulation, some areas of the ocean basins are covered by hundreds of meters of sediment.

The outer skin of the Earth is divided into a relatively small number of rigid plates, which are approximately 60 miles (100 kilometers) thick and bounded by midocean ridges and deep-sea trenches. The direction of the motion between any two plates determines the nature of the plate boundary. Plates separated by a midocean ridge are clearly moving apart, while plates separated by deep-ocean trenches and active mountain chains, or both, are moving toward each other. Where plates are moving past each other without the creation or destruction of crustal material, they are separated by major fault zones, as in the Caribbean Basin.

The movement of plates is also indicated by volcanic island chains, such as the Hawaiian islands, which result from the movement of the ocean plate over a fixed 'hot spot' in the mantle. Passing northeastward along the Hawaiian island chain, the islands become progressively older, with the oldest volcanoes no longer reaching the surface of the sea, but continuing the chain in the form of a line of submerged guyots, or sea mounts.

The magnetic record

New rocks are formed at ocean ridge crests, the magnetic particles are aligned parallel to the Earth's magnetic field.

The Earth's magnetic field has changed its polarity in the past. Any new rocks are magnetized in the new direction.

A symmetrical arrangement of magnetized strips was discovered south of Iceland. The pattern of reversals correlates with those in the Pacific and with the sequence over the past 10 million years.

CONTINENTAL DRIFT

The way in which the present day continents can be fit together into a single landmass like the pieces of a giant jigsaw puzzle led to the idea that the continents were once joined together. This idea is not new and as early as 1858, maps were published that depicted the continents welded into a single, large landmass, surrounded by a vast ocean. Subsequently, the evidence of the similarity in both fossil type and age of geological formations on both sides of the Atlantic led Alfred Wegener, in 1915, to publish a series of reconstructions of the continents.

However, it was not until the 1960s, when the evidence of paleomagnetism (the record of the Earth's magnetic history) in the ocean crust became available and mechanisms to explain the process of continental drift were postulated, that the theory of continental drift became accepted. Around this time, Paleomagnetism studies showed that ocean crust in the vicinity of the midocean ridges was younger than the crust that was further away leading to our present understanding of the process of seafloor spreading.

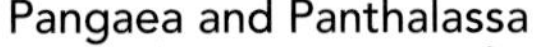

Pangaea and Panthalassa

The Earth's crust consists of 12 plates that float on the surface of the denser molten material beneath. These plates are constantly moving and through geological time have collided or separated as the molten material beneath the crust circulates. As the plates with their continental landmasses

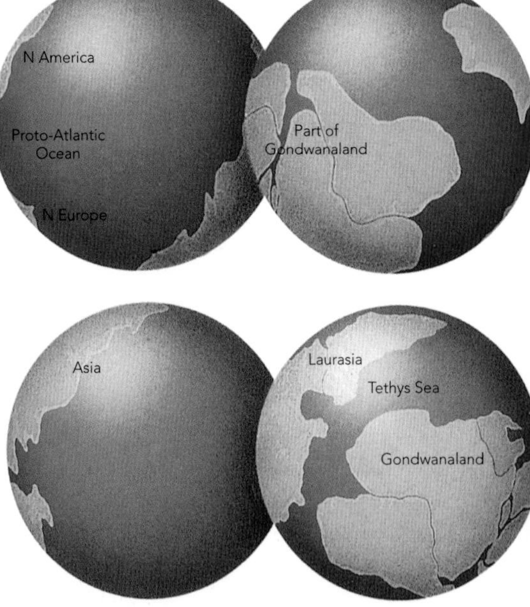

► ***The Cambrian world*** *would be unrecognizable today. Gondwanaland existed as a single landmass that would not split into Australia, India, South America, Africa and Arabia for another 400 million years. North America, northern Europe and most of Asia were separate continents.*

► ***In the Carboniferous,*** *North America and Europe had closed the proto-Atlantic Ocean, forming the landmass of Laurasia with the Caledonian Mountains between. Laurasia was separated from Gondwanaland by the Tethys Sea and the South Pole lay in South America.*

◀ ***Antonio Snider-Pelligrini*** *was the first to reassemble the continents according to geological as well as geometrical evidence. He published the maps on the left in his book,* La Création et ses Mystères Dévoilés, *(1858). His ideas were deemed too farfetched, and were forgotten for more than 50 years.*

collided, oceans have closed, and as they separated, new seas were formed.

During the Cambrian period (590–505 million years ago), most of today's southern hemisphere landmasses were fused into a supercontinent known as Gondwanaland. North America, northern Europe and most of Asia (later to become Laurasia) remained separate landmasses. By the Carboniferous period (350 million years ago), North America and northern Europe had collided and, where these two continental plates met, the Caledonian mountain range was formed. The movement of southern Europe toward northern Europe raised the Hercynian Mountains, while the collision of Asia and Europe created the Urals.

For a time, the newly formed landmass was separated from the larger continent of Gondwanaland, but by the Permian period (280–225 million years ago), the continents fuesed together to create the supercontinent of Pangaea. A single superocean, known as Panthalassa, surrounded the mighty Pangaea.

Subsequently, between 200–180 million years ago, Pangaea began to split into the northern continent of Laurasia and a southern continent of Gondwanaland, and a new shallow sea, the Tethys Sea, was formed between them. Thirty million years later, Gondwanaland began to separate, forming the Indian and South Atlantic Oceans, the latter being 100 million years old and dividing the southern continent into South America

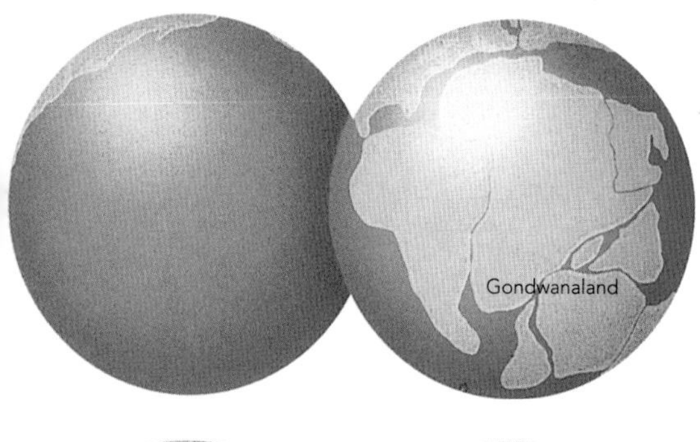

◀ ***The Permian*** *was the time of Pangaea and Panthalassa, with the land forming a single supercontinent. The movement together of Gondwanaland and Laurasia pushed up the Ural and Hercyian mountain ranges.*

◀ ***In the early Cenozoic,*** *flowering plants had appeared and mammals had replaced the reptile. Pangaea had broken up and the Atlantic had formed. Antarctica and Australia were still together, but India was drifting northward to collide with Asia and produce the Himalayas.*

and Africa. The Indian Ocean was formed when Africa split away from a combined landmass of what is now Australia and Antarctica.

Soon after its split from Pangaea, Laurasia began to break apart, forming the North Atlantic, although the North American and European continents were joined in the North as little as 65 million years ago. The Pacific Ocean was separated from the Atlantic Ocean when the North and South American continents joined, while Australia moved away from Antarctica and drifted northward toward its present-day position. When Greenland separated from North America, the encirclement of the Arctic Ocean was complete.

Ocean basins

The ocean basins have been formed, lost and reformed many times. Six hundred million years ago, for example, an ocean, the proto-Atlantic, not unlike that of the present day, covered the North Atlantic region. Around 500 million years ago, this proto-Atlantic had started to close and was flanked by Andean-type mountain ranges. About 100 million years later, it had completely closed and the movement of the continents toward one another formed the Caledonian Mountains. The present-day North Atlantic has only been in existence for the last 20 million years, and the split, which formed this ocean basin, occurred roughly along the line of closure of its predecessor, the proto-Atlantic.

If a new area of upwelling in the asthenosphere starts beneath a continental area, this sets up tensions in the continental crust. The tension causes the crust to split and form the classic rift valleys such as those of eastern Africa. As the two parts of the continent move away from one another, a new sea is formed, along the center line of which a midocean ridge develops and new ocean crust is formed. Today, the Red Sea represents an early stage in the formation of a new sea, as Africa drifts away from Arabia. Widening at the rate of about 0.8 inches (2 centimeters) per year, the Red Sea forms a new link between the Mediterranean and the Indian Ocean. On the other side of the Arabian peninsula, the Persian Gulf began to form only between 3 and 5 million years ago.

Mountains and trenches

When two continental plates collide, the continental crust buckles and is forced both upward and deeper into the underlying lithosphere. The result is often the formation of mountain ranges, such as the Alps and Himalayas. The

***◄▼ The configuration** of continents today reflects the movements of the continental plates over the past few hundred million years. Similar geological features on separate continents allow geologists to reconstruct the past positions of the land and oceans.*

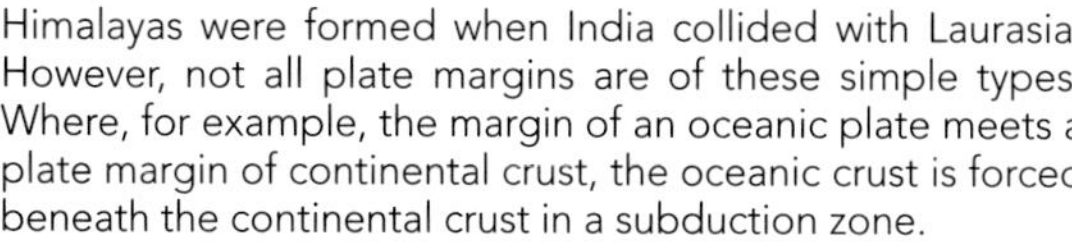

Himalayas were formed when India collided with Laurasia. However, not all plate margins are of these simple types. Where, for example, the margin of an oceanic plate meets a plate margin of continental crust, the oceanic crust is forced beneath the continental crust in a subduction zone.

An example of a subduction zone can be found off the west coast of Latin America. It takes the form of a deep ocean trench, where the denser ocean crust is being forced downward. In such situations, the less dense continental crust is folded and uplifted to form extensive mountain ranges such as the Andes in South America, the Rocky Mountains in North America, and (during a geologically earlier phase of continental drift), the Appalachians.

The continental shelves along the ocean margins are formed by compression, with sediments being actively incorporated into them as the oceanic crust slides beneath the continental plate. In some more restricted areas, such as the coast of California, complex continental borderlands with a mixture of plateaus and basins are found. These represent the fragmented pieces of continental plates moved together by different processes.

In some ocean basins, such as the Pacific, two plate margins, both formed of ocean crust, move toward each other, and one is forced beneath the other. A deep ocean trench, where one plate is driving down into the crust, characterizes such destructive plate margins, and an arc of volcanic islands forms on the other plate.

Continental history

Alfred Wegener, a German geophysicist, astronomer and meteorologist, published a series of reconstructions of the continents in 1915 (translated in 1924 as *Origin of Continents and Oceans*). His evidence was based on the similarity in fossil type and age of geological formations on both sides of the Atlantic. His work was not recognized until the 1960s. Although he was mistaken about the speed of drift, his maps are still acceptable today.

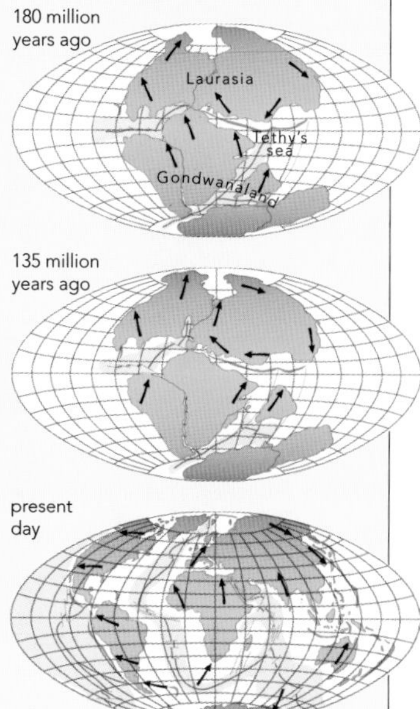

Shorelines vary in their form since they are shaped by the action of both wind and waves. The movement of material by currents in near-shore waters may result in the deposition of sands, gravels or larger material, such as boulders, depending on the power of the waves. In other areas, where sedimentary material is scarce and the waves move material away from the shoreline, erosion dominates the coast. Eroding shorelines tend to occur in areas of high wave energy, while depositional shorelines are characteristic of sheltered coasts.

▲ ***A section*** *of the Giant's Causeway, a coastal formation of basalt columns that extends into the sea off the tip of Northern Ireland.*

Eroding coasts

Contrary to popular belief, water itself does not erode the land, it is the particles of sand, pebbles and rocks caught up in the waves crashing onto the beach that cause erosion. Where a shoreline has a steep slope, the wave energy breaks directly onto the shore and erosion is high; where the slope is low, the energy of the waves is dissipated over long distances and the erosive force of the wave is reduced.

Cliffs, and dramatic rocky features, such as the Giant's Causeway off the Irish coast, often characterize eroding shorelines. Although the height of cliffs facing the sea is largely determined by the height of the land behind them, their form is dependent on a wide variety of factors, including the properties of the rocks and sediments of which they are composed, the topography of the landscape into which they are cut, and the geological history, including sea level and tectonic changes. Great variations are possible but features, such as promontories, blowholes, arches and sea stacks, are frequently formed in well-jointed rocks and give rise to some of the most distinctive coastal scenery.

In general the nature of eroding shorelines reflects the ability of the rock to resist the erosive forces of the waves and the suspended materials thrown by the water against the shore. Most erosion takes place along weaknesses in the rock, such as faults, joints, bedding planes and layers of softer materials interspersed between harder bands.

Coastal evolution

When a wave approaches the shore it is slowed down in shallow areas and on meeting an irregular coastline, is turned along the shore. Most of the wave's energy is expended against projecting headlands and as a result these are eroded. Eventually the headlands may completely erode and a straight shoreline will evolve.

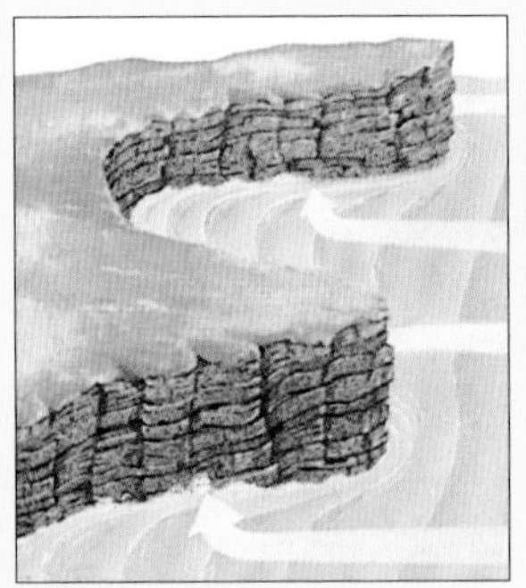

▲ ***A headland cliff*** *erodes first along joints or bedding planes, which then enlarge to form deep clefts.*

Erosion occurs in a relatively narrow band centered on mean sea level. The width of the band is narrow in areas with a small tidal range and wider where the range is greater. The restricted width of the zone of active erosion means that the removal of material occurs at the foot of the cliff and results in collapse of materials as they are undercut by wave action. In areas of soft rock cliffs, the rate of erosion may be limited only by the rate at which longshore currents remove the slumped material from the base of the cliff where it forms a protective beach. Along hard rock formations, the undercutting may result in only infrequent collapse of the cliff face above, and shallow sea caves or overhangs may develop. Many raised limestone islands in the Pacific have distinctive notches around their coasts that mark the present sea level. Some occur below and others above present sea level, marking periods of high and low sea level during the ice ages.

In rocks such as sandstones, which have even bedding planes (surfaces that separate different layers of rock), erosion occurs along the planes; softer strata are worn away faster, leaving a ragged profile to the cliff face. In fine-grained chalks where bedding planes are not well defined, smooth, sheer cliff faces develop, while in basalt, erosion takes place along the joints. Basalt cliffs, therefore, appear to have been built of individual blocks.

Beaches on eroding shorelines

Along any eroding shoreline, a beach of eroded materials will be formed as the cliff face collapses into the sea. These erosional products may be either removed by currents and

▲ ***A vertical cleft*** *subjected to constant erosion will eventually enlarge to form a cave.*

▲ ***When air*** *in a cave is repeatedly compressed by waves, the pressure forces a hole in the roof.*

▲ ***Deepening of caves*** *on opposite sides of a headland causes them to meet, forming a rock arch.*

▲ ***A natural arch*** *will continue to widen until its lintel collapses, leaving an isolated sea stack.*

deposited elsewhere, or deposited as an offshore terrace or a wave-cut platform beneath the erosional face of the cliff.

The beaches found at the foot of cliffs are transitory, often changing their form with the season as changing wind and wave patterns alter the rate of removal of the material. In general, the finer materials are removed more rapidly so that pebbles, cobbles and even boulders tend to dominate such beaches. Most of the erosional materials that form these beaches are removed offshore to deeper waters or moved along the coastline to more sheltered areas where depositional coastal formations occur.

Along some coastlines, such as those of the eastern Mediterranean, pocket beaches are found at the head of small inlets. Such beaches are generally surrounded by high cliffs, and these coastal landforms are highly susceptible to changes in sea level or wave patterns, which result in the sediments being lost offshore.

In the case of many Mediterranean islands, which have been formed by uplift along fault lines, cliffs descend vertically into the sea and erosion rates may be quite slow since the tidal range is small. In contrast, along low-lying coastlines where relative sea level is rising and the materials are

Coastal erosion

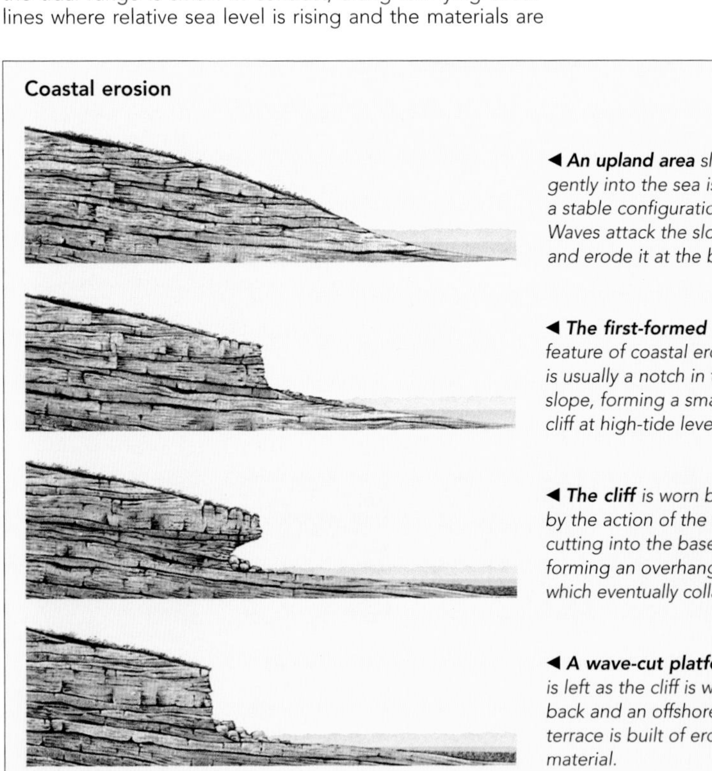

◀ ***An upland area*** *sloping gently into the sea is not a stable configuration. Waves attack the slope and erode it at the base.*

◀ ***The first-formed*** *feature of coastal erosion is usually a notch in the slope, forming a small cliff at high-tide level.*

◀ ***The cliff*** *is worn back by the action of the waves cutting into the base, forming an overhang which eventually collapses.*

◀ ***A wave-cut platform*** *is left as the cliff is worn back and an offshore terrace is built of eroded material.*

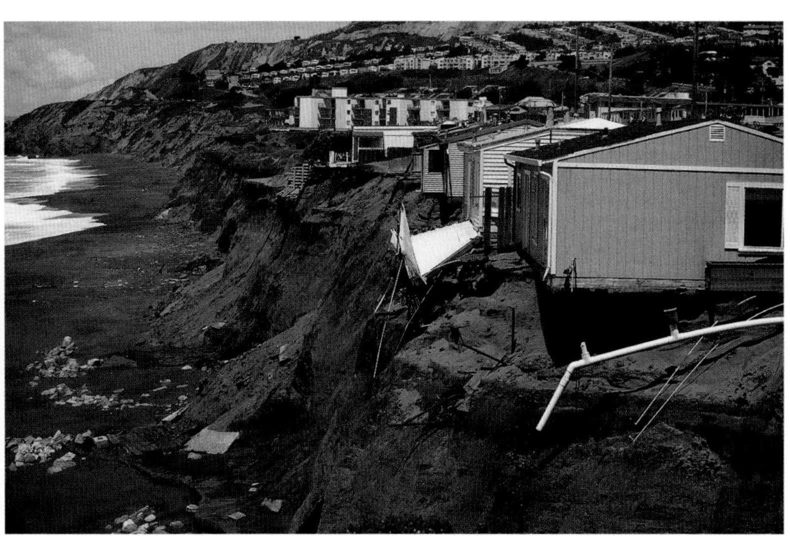

▲ ***Pacific Ocean eroding*** *the beach under seafront houses, at Pacifica, California, USA. The demand for ocean-view housing has encouraged development on fragile coastlines. The action of waves and weather on the world's coasts means that they are constantly changing shape. Seafront risk destruction from crumbling foundations and assist in the destabilization of the very land on which they are built.*

soft and unconsolidated, erosion rates may be extremely high, as much as several meters a year. Often the erosion occurs in short periods associated with storms, resulting in the development of an erosional scree slope before the materials are gradually removed.

Life on eroding shores

Erosional shores are generally inhospitable environments for animals and plants. Since they are characteristically high wave-energy environments, the animals and plants must be able to either attach themselves firmly to the rocks or hide away in the cracks. Where the rates of erosion are high, the diversity of the community of animals and plants tends to be limited since the shore does not provide a suitably stable base for attachment. On hard rock, erosion-resistant shores, the diversity of organisms tends to be greater and the zonation of the animal and plant communities is well marked.

Above the high-tide level in the splash zone are found species which are resistant to desiccation, while soft-bodied animals generally occur lower down in the intertidal zone where they are active only when covered by the tide. Above the high-tide line occur a wide variety of lichens and some blue-green algae, which are replaced lower down by the typical sequence of green, brown and red seaweeds. On rocky shores where the surface is smooth, animals tend to exist lower in the intertidal zone than on shores where cracks and fissures are present, which can provide moist hiding places. Thus the nature of the material forming the shoreline not only affects the rate of erosion but also the distribution of the organisms that live there.

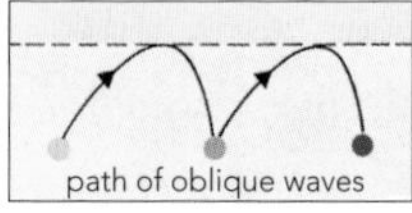

▲ ***Longshore drift*** *is caused by waves meeting the beach at an angle.*

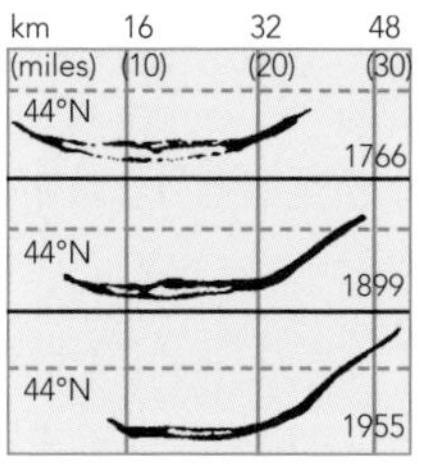

▲ ***Sand-formed features*** *may be moved great distances by longshore drift. In 200 years, Sable Island, off Nova Scotia, eastern Canada, traveled 9 miles (14 km) eastward.*

Deposition on coastlines occurs from the buildup of sediments, derived from the weathering of exposed rock surfaces and erosion from land. Eroded materials, including soil, are moved by rain into the river systems of the world where they may remain for some time in flood plains and river levees before being discharged to coastal waters.

Deltas and estuaries

Most of the major rivers of the world enter the sea via large, deltaic systems, constructed of sand and fine muds deposited at the delta margin as a result of the reduced speed of water movement when the river meets the ocean. Coarser sediments settle out first, while fine materials settle further offshore. The movement of the tides reworks the sediments into flats, which may be exposed at low tide.

Sediments from the land are often moved along the coast by the process of longshore drift. Thus beaches may be dependent on a source of sediment that is a considerable distance away. When the supply of sediment from a river is cut off (by dam construction, for example), beach erosion can occur down current.

Sediment types

The nature of the sediments found along shorelines reflects their source. In the temperate region, beaches are normally formed of materials of mineral rather than organic origin. In contrast, along many tropical shorelines, entire beaches are comprised of bioclastic sand, formed by the weathering of the skeletons of corals, mollusks and other marine organisms. In many atoll environments, much of the calcium carbonate found on beaches was originally laid down in the living tissues

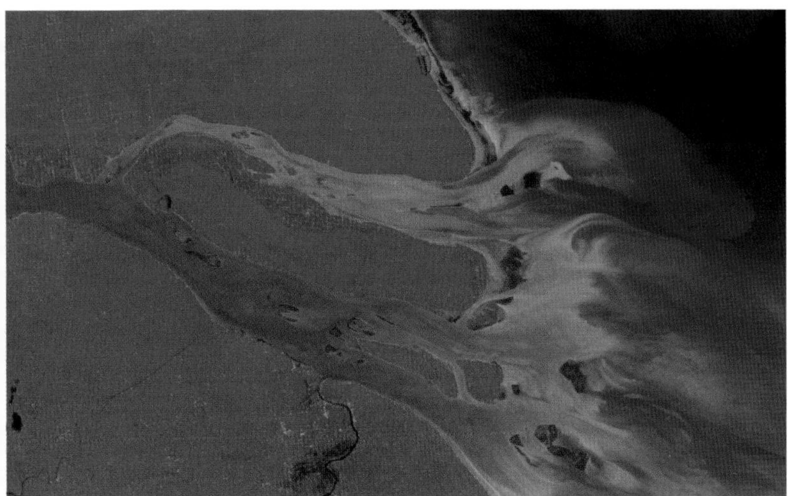

▲ ***The swirls** of blue and beige in this Landsat picture of the Yangtze (Chiang Jiang) river delta in eastern China show that the delta is continuing to grow seaward as sediments are brought by the river from the inland areas of the catchment.*

◀ ***View of a sandy coastline** showing marram grass,* Ammophila arenaria, *growing on the dunes. Marram grass is important in coastal ecology, as it is one of the most common species to stabilize sand dunes. It is frequently planted for exactly this purpose.*

of corals and coralline algae. These white beaches contrast sharply with the black sand beaches formed of weathered volcanic materials on oceanic islands and island arcs.

The action of the waves and currents not only reduces the size of individual particles through mechanical abrasion, but also sorts the material into different sizes. Toward the landward side of a beach, the particle size is larger than lower in the intertidal zone, while the finest sediments are found offshore as subtidal muds.

Life in soft shores

The sediments on the sea bottom are constantly moved by tidal and other currents – ripple marks and other small-scale features of the surface reflect this movement. This disturbance is greatest in the tidal and subtidal reaches of the shore, and as a result the surface of soft shorelines is home to relatively few animals and plants.

Most organisms on soft shores have adaptations for burrowing and spend much of their life below the surface, emerging to feed and mate when water conditions are suitable, or merely extending the feeding apparatus above the surface while the body of the animal remains in its protective burrow.

At low tide a mud flat may appear lifeless and barren, but this appearance is misleading. Buried beneath the sediments lie a multitude of animals ranging from small microscopic forms, which move between individual sand grains, to large burrowing clams and lugworms eagerly sought by seabirds that forage in such areas at low tide. The distribution of these animals reflects not only the influence of the tidal range but also the size of sediment particles with different species inhabiting areas with different types of sand, mud and clays.

MAN-MADE SHORES

At the present time, around 60 percent of the world's total population lives within 40 miles (60 kilometers) of the coast, and in many countries this proportion is much higher. Historically, large-population centers tended to develop in coastal locations, deriving much of their economic viability from international trade and commerce. Coastal tourism continues to grow worldwide, and today over two-thirds of the world's cities with populations greater than 1 million people are located on the coast.

▶ ***The Thames Barrier**, London, England. If a high surge coincides with a high 'spring' tide, there could be a danger of flooding along most of the tidal Thames. The barrier is a series of movable gates positioned end-to-end across the river*

Land reclamation

As a consequence of these trends, the demand for highly priced coastal land near major ports, harbors or tourist developments has resulted in expensive reclamation schemes. In cases where the area of land is limited, such as on islands, new land is formed by pumping lagoon sand onto the surface of neighboring coral reefs. Male, capital city of the Republic of the Maldives, for example, occupies the entire surface of a coral island, 5610 feet (1700 meters) long and 2310 feet (700 meters) wide, over half of which was artificially constructed. As a result, the island occupies virtually the entire surface of the coral reef platform, and the seaward edge of the island is only some 100 feet (100 feet) from the edge of the reef platform. To protect the island, a sea wall was constructed around the entire perimeter to a height of about 7 feet (2 meters) above sea level. A breakwater has now been constructed on the seaward side of the island at a cost of US$40,000 per foot. Land reclamation and the subsequent need to protect the investment inevitably results in further investment in protection. This example is replicated in different forms all around the world.

▼ ***Aerial view of Male**, capital of the Maldives. Male is located on a coral bank and protected by a long sea wall built around the perimeter.*

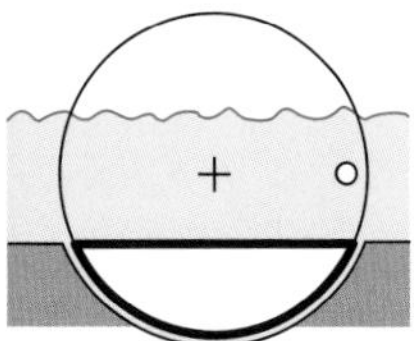

A Gate in open position (flush with river bed)

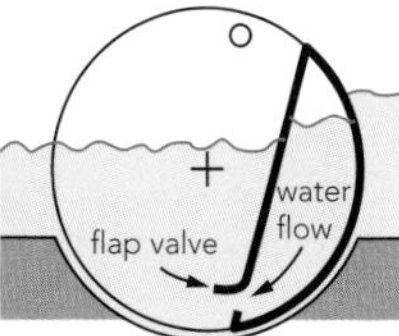
water flow
flap valve

B Gate rising

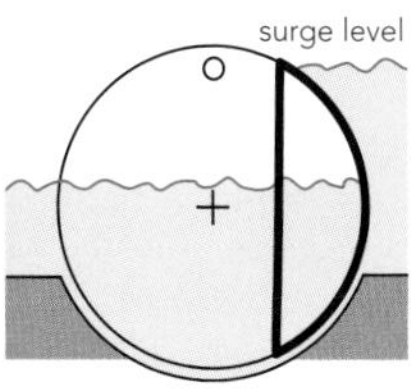
surge level

C Gate in flood-control position

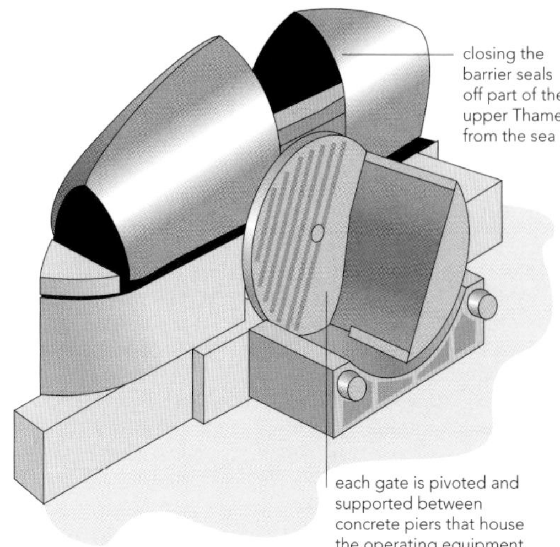
closing the barrier seals off part of the upper Thames from the sea
each gate is pivoted and supported between concrete piers that house the operating equipment

Coastal protection

Under natural conditions, shorelines change, sediments move on- and offshore, and in many areas are moved along a coastline. The construction of hard structures at right angles to the shore interrupts this flow of sediment, resulting in accretion on the updrift side of the groine and enhanced erosion on the downdrift side.

The dynamic nature of natural shorelines is an inconvenience to many human activities, and considerable effort is expended in trying to hold the shoreline in a constant position. When erosion commences, and buildings or roads are threatened, beaches may be armored, sea walls built and groines constructed. Loss of sand from tourist beaches has resulted in expensive beach replenishment schemes in which sand is pumped from deeper waters offshore onto the beach. Like dredging of harbors, this activity is an ongoing expense.

Coastlines worldwide are being dramatically altered, and in some countries, such as the Netherlands, the whole shoreline is artificial. What was once a dynamic boundary between land and sea, subject to erosion and/or accretion under different conditions, has been fixed at one point in space. Much of the Netherlands is actually below sea level, having been reclaimed through construction of dykes and pumping of water out of the enclosed polder. Maintaining this dry land requires constant and continuous pumping of water into the sea from behind the protective dykes. The protective dykes themselves must be constantly maintained if the sea is not to encroach on what is now intensively used farmland and densely populated areas. Maintaining artificial coastlines will become increasingly expensive in the future.

CHANGING LEVELS OF THE SEA

Anyone who has lived near the sea or spent time on the seashore is aware of the daily rhythm of the tides. The tides can change the meeting point of sea and land by as much as 50 feet (15 meters) vertically and thousands of meters horizontally. In addition to the daily variation in sea level resulting from the tidal cycle, the position of high and low water varies between spring and neap tides depending on the phase of the lunar cycle. Unusually high water levels may occur when onshore winds occur at the same time as high tides. During such events, seawater may penetrate far inland along freshwater courses and inundate coastal land.

The more subtle and gradual changes in sea level are less obvious. Some of these changes represent long-term trends, such as the apparent rise in global sea level of around 1/16 inch (1.5 millimeters) per year, which has occurred over the last 100 years. In contrast, other changes represent shorter-term responses to major shifts in ocean currents, such as the lowering of sea level in the western Pacific by as much as 6 inches (14 centimeters) during El Niño years (*see* page 58).

Global changes in sea level

Global sea level is affected by a variety of factors, including the volume of water in the ocean basins, thermal expansion of the surface layers of the ocean, and changes in depth of the ocean basins. During the Pleistocene period, the large volumes of water alternately locked into icecaps on land or released as water to the ocean basins, resulted in changes in sea level of up to 400 feet (120 meters).

Local changes

On a local scale, relative sea level may be affected by a wide variety of phenomena. In the Mediterranean, tectonic move-

▼ ***Venice** lies on a small, fish-shaped island in the Laguna Veneta at the northwest corner of the Adriatic Sea, northeast Italy. For centuries, the low-lying city has successfully coped with a 0.9-meter (3-foot) tidal range experienced at this end of the Adriatic Sea, and the series of barrier islands has offered some protection from storm waves. Today, a combination of both regional land subsidence and recent rises in sea level pose a significant threat to this historic city*

◀ ***The collapse*** *of the Kolka Glacier on Mount Kazbek, southern Russia, August 13, 2002. Although scientists have predicted the possibility of large glacial collapses as the climate warms, no one predicted that tragedy would strike the mountain village of Karmadon so soon.*

ments have resulted in the submergence of some Roman and Bronze Age ports and the uplift of others which are no longer at sea level.

In some areas, coasts are sinking or rising more gradually. The southern coast of England, for example, is slowly sinking, while that of northern Scotland is rising. During the glacial periods, the northern half of these islands was covered by extensive ice sheets, the weight of which depressed the Earth's crust, causing the north to sink and the south to rise. Once the ice melted, the crust readjusts, causing the south to sink and coasts to erode, and the north to rise, resulting in raised beaches above the high-tide line.

▼ ***This map*** *shows the dramatic effects of possible sea-level rise on the state of Florida, southeast United States. Sea-level rise is one of the effects of global warming. Sea-level rise on this scale would affect more than 7 million residents of Florida.*

Human-induced changes

Human actions can also cause adjustment of the land surface. Extraction of groundwater, oil and gas from coastal and near-shore areas causes problems of rising relative sea level in many coastal areas around the world. In Bangkok extraction of groundwater caused compaction of the sediments and lowering of the land surface.

A more recent concern is the potential effect of global warming which may cause the further melting of land-locked ice. In addition to this melting, increased surface temperatures will result in expansion of the surface layers of the ocean, which may in turn cause a rise in relative sea level of up to 3.5 inches (9 cm) by the year 2100, although this is not accepted by all scientists.

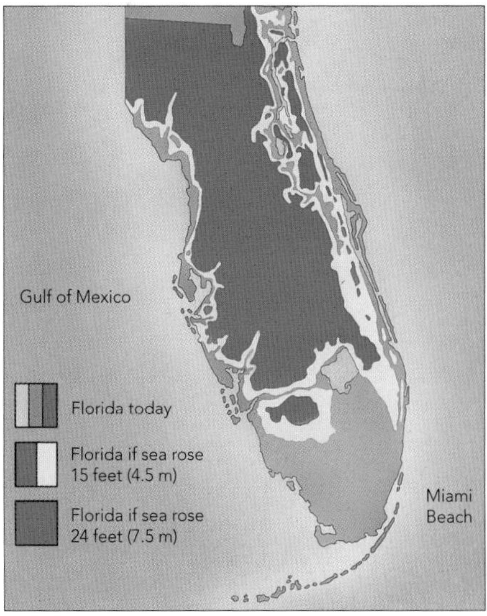

Surrounding the margins of most of the world's continents are terraces with a coastal plain lying above sea level and a continental shelf that extends out below the ocean surface. The outer edge of the continental shelf is at a relatively constant depth of 500 feet (150 meters). The constant depth reflects the fact that the continental terraces were formed around 15,000 to 20,000 years ago when the sea level was lower than today. The outer edge is also the real limit of the continental plates, separating them from the oceanic crust of the deep ocean floor, which lies at a depth of around 12,000 feet (3500 meters).

Continental margins

Separating the continental shelf and the deep ocean floor is a steep continental slope having a grade of between three and 20 degrees. Passive, Atlantic-type ocean margins, formed by the splitting apart of continental blocks, are marked by a gradual continental rise running up to the foot of the continental slope, and wide continental shelves backed by extensive coastal plains.

In contrast, active, Pacific-type ocean margins are areas of high earthquake and volcanic activity. The continental shelves of such margins tend to be narrow, and they are separated from the ocean floor by deep ocean trenches that mark the point at which one plate is sliding beneath another. High mountain ranges rather than a broad coastal plain often back the neighboring land.

▼ ***The continental shelf*** *is the submerged edge of a continental plate. The surface of the continental shelf off eastern North America shows typical features of a past land surface. Elephant teeth and freshwater peats have been found on the ocean bed.*

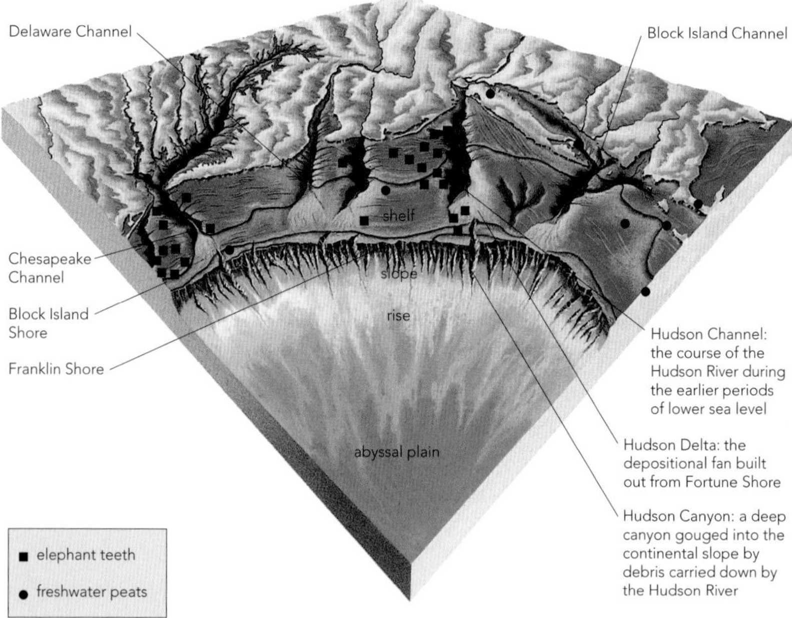

Profile of the continental margin

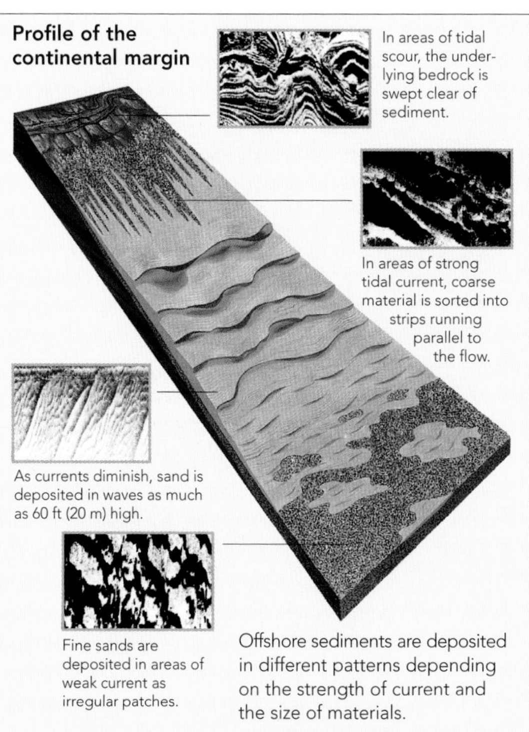

In areas of tidal scour, the underlying bedrock is swept clear of sediment.
In areas of strong tidal current, coarse material is sorted into strips running parallel to the flow.
As currents diminish, sand is deposited in waves as much as 60 ft (20 m) high.
Fine sands are deposited in areas of weak current as irregular patches.

Offshore sediments are deposited in different patterns depending on the strength of current and the size of materials.

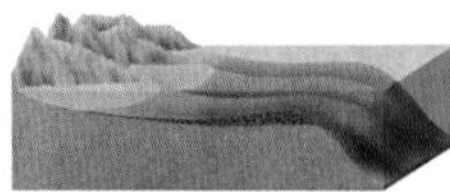

Some continental shelves have deep gorges, caused by river erosion before the land was submerged or by ocean currents.

Other regions have shelves with a gentle relief, often with sandy ridges and barriers.

In high latitudes, floating ice wears the shelf smooth.

In clear tropical seas, a smooth shelf may be rimmed with a coral barrier reef.

When strong, offshore currents wash sediment away, shelves can start with a gradual slope, but then drop suddenly.

Shelves can also be affected by faulting.

Turbidity currents

On the surface of the continental shelves, river valleys, formed at times of lower sea level, drain toward the edge of the continental slope. Such drainage systems continue to be eroded and may be cut deeper by turbidity currents.

Turbidity currents are dense mixtures of sediment and water that flow rapidly down the continental slope. They may be initiated by slumping of loosely consolidated sediments at the head of drainage systems near the continental slope. The speed of such currents was dramatically illustrated in 1929 when an earthquake shook the Grand Banks area off the east coast of North America. Submarine cables were broken by the current, and the timing of the successive breaks provided an estimate of the speed of the current at 55 knots. Cables snapped as far as 300 miles (480 kilometers) away from the source of the shock.

The coarser debris carried to the foot of the continental slope by turbidity currents may be built up into screes at the head of the continental rise. Finer particles may be retained in suspension just above the seafloor. Large fans of material mark the position of canyons leading from the continental shelf to the deeper ocean floor.

The sediments of the deep ocean floor accumulate through passive sinking of materials from above and vary in both composition and origin. Bottom sediments have two origins: bioclastic materials, which are formed from the dead remains of animals and plants that lived in the surface waters of the ocean; and terrigenous materials, which are ultimately derived from erosion of the land surface.

Land-based sediments

Most land-based materials are deposited on the continental shelves, with larger material being deposited inshore and finer materials being carried further away. Red clays, found in the deepest ocean basins, are composed of the finest clay particles and are the major type of deep-sea sediment derived from land. The clay particles remain in suspension more easily and are moved farther away from the continental margins than sand and larger particles. Red clays have been found in ocean basins at depths of more than 3300 feet (1000 meters).

Deep-sea fans of material, some of them millions of years old, are located off the continental shelves in front of large rivers. Of these the Ganges fan is perhaps the most striking. It extends as a cone beyond the continental shelf for a distance of around 1500 miles (2500 kilometers) and ends at a depth of around 16,500 feet (5000 meters).

Other land-based material found in some areas of the deep-ocean floor includes volcanic ash, which can fall into the ocean thousands of kilometers from its site of eruption.

▼ ***Unconsoldiated*** *sediments on the ocean floor are classified according to the nature of their principal constituent. Distance from landmasses, the nature of the winds and currents, the surface water temperature, and water depth determine the type of constituent. The major types are: the terrigeneous deposits, which are debris derived from the weathering of continents; the red (or brown) clays that form inorganic sediments; and the globigerina, radiolaria, pteropod, and diatomaceous oozes, which are formed by the accumulation of shells of deceased planktonic animals.*

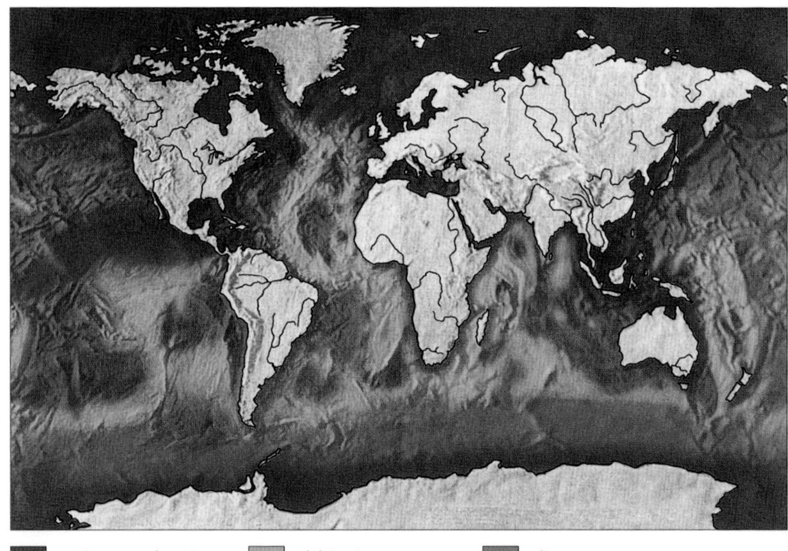

Midocean ridges

The shape of a midocean ridge reflects the rate at which it is spreading. The axis of a slow-spreading ridge, such as the Mid-Atlantic Ridge, lies in a deep rift valley surrounded by faulted mountains. A fast-spreading ridge, such as the East Pacific Rise, stands higher than the surrounding seabed terrain.

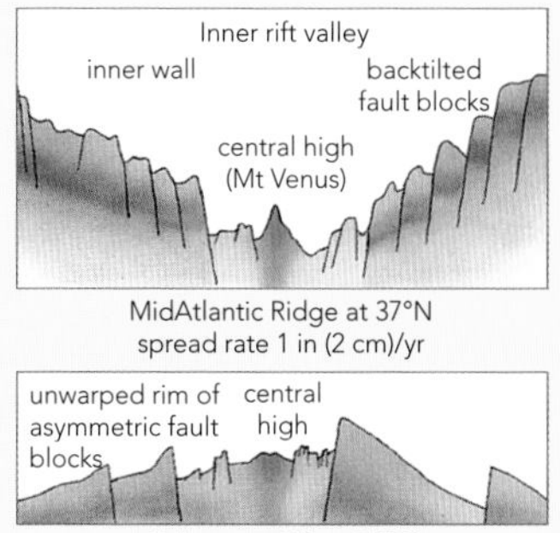

MidAtlantic Ridge at 37°N
spread rate 1 in (2 cm)/yr

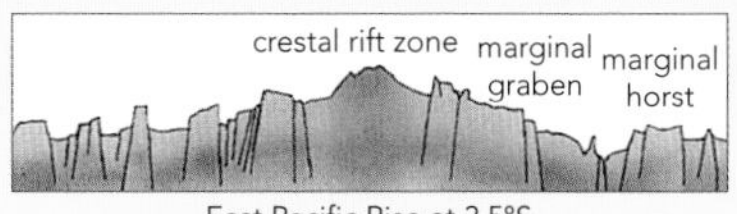

East Pacific Rise at 3.5°S
spread rate 6 in (15 cm)/yr

Galapagos Rift at 86°W
spread rate 3 in (7 cm)/yr

Bioclastic sediments

Found at depths of 3000 feet (1000 meters) or more are the bioclastic oozes, or pelagic deposits. Calcareous sediments are composed of the chalky remains of foraminifera and pteropods, whereas silaceous sediments are formed of the shells of radiolarians and diatoms, the single-celled phytoplankton characteristic of the surface waters of the ocean.

The thickness of deep-ocean sediment varies. Newly formed crust, close to the midocean ridges, has little or no sediment cover, and the shape of the newly extruded volcanic lava is easily seen. Moving away from the ridge toward the ocean margin, the depth of sediment increases. At around 2–3 miles (3–5 kilometers) from the ridge, the lava flows and tubes and volcanic cones are partially hidden, and at distances of 6–8 miles (10–13 kilometers), the ocean floor is a featureless surface, interrupted only by the tops of the largest lava masses.

In addition to land-based sediments and sediments of organic origin, metalliferous muds may form as a result of precipitation of metals in solution when hot molten materials from the underlying crust come into contact with the cooler ocean water.

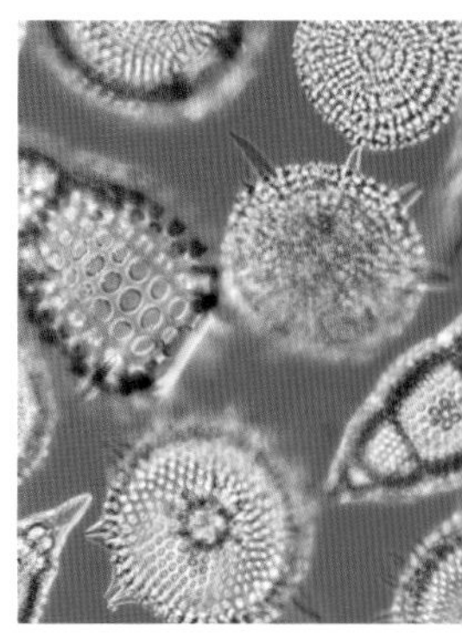

▲ ***Diatoms** help to form deep-sea oozes.*

Shifting ocean sands

Earthquakes, volcanic activity and deep-water currents move sediments from one area to another, resulting in rapid accumulation in some areas and scouring of sediments in others. Where soft, unconsolidated sediments are deposited in areas of tectonic activity, slumping and sliding of sediments is common, but in areas where bottom or strong midwater currents meet submarine features (such as submerged seamounts or ocean ridges), scouring of the bottom may occur. Acoustic side-scan sonar has revealed that vast transverse and crescent-shaped sand waves move across the ocean floor in much the same way as similar types of sand features move across continental deserts – such as the Sahara in North Africa.

WATER: THE UNIQUE COMPOUND

Despite its commonplace occurrence and familiarity, water is a unique compound and one on which all life on Earth depends. Its physical and chemical properties shape the face of the planet, determine the nature of our climate, and support all life on land and in the oceans.

The physical properties of water are distinctly unusual. Most substances expand when heated and contract when cooled, and although this is also true of water over part of its temperature range, water does not behave in this way over the full range of temperatures experienced on Earth. As water is cooled, it continues to contract until it reaches temperatures of about 40°F (4°C). Below this temperature, however, water freezes and expands. As it freezes, water increases in volume by around 9 percent. As a consequence, frozen water, ice, is less dense than liquid water and so it floats on the surface of the sea. In this way, ice forms an insulating layer that prevents further cooling of the deeper water layers. If this did not occur, and water continued to contract as it froze, then frozen surface water would sink and the polar seas would eventually become solid ice.

► ***A huge wave** breaks as surf on the shore, producing white water at the crest. The movement of water across the oceans is important to the regulation of climate.*

▼ ***The water** (or hydrological) cycle is the process of water exchange between oceans, atmosphere and land. Water in the oceans evaporates, creating water vapor in the atmosphere that may condense and fall back to the ocean. Winds carry water vapor over the land, and evapotranspiration from rivers, lakes, the land and plants adds to atmospheric moisture. Precipitation is absorbed by the vegetation or flows on the surface in rivers and streams. Some infiltrates the ground via porous rock and may drain into lakes or rivers; in other areas lake water drains into the groundwater. Groundwater may be important for vegetation, while in coastal areas it is a component of the drainage to coastal waters.*

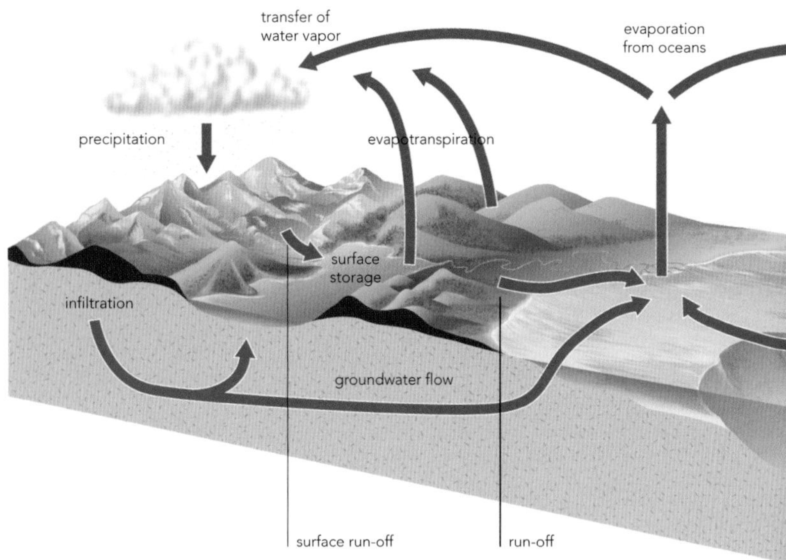

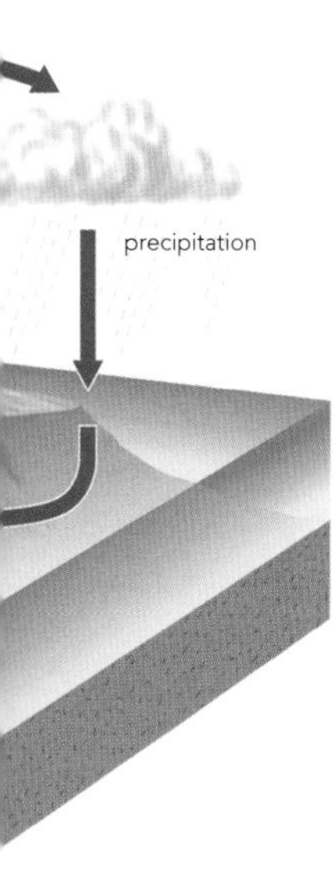

The hydrological cycle

Not only does water occur as a solid and a liquid, but also water vapor is a major component of the Earth's atmosphere. When heated, water evaporates to become water vapor, which passes into the atmosphere where it forms clouds. The water vapor eventually condenses, falling as rain or snow, often at considerable distances from the point of evaporation. This transformation of water from liquid to gas and back to liquid, and its passage from oceans and surface waters to atmosphere and land, is known as the hydrological cycle. This cycle is fundamental to terrestrial life, since without a source of freshwater, life on land would not exist.

Every year, around 80,000 cubic miles (330,000 cubic kilometers) of seawater evaporates from the surface of the oceans and enters the atmosphere as water vapor. Of this total, some 24,000 cubic miles (100,000 cubic kilometers) falls as rain, sleet and snow on the surface of the land. The remainder falls directly back into the oceans. Rainfall is made slightly acidic by carbon dioxide (CO_2) in the atmosphere, and this acidic solution attacks the rocks on which it falls. The chemical action, combined with changes in temperature and the physical abrasion of wind and rain, gradually erodes the surface of the land. Erosion products are then carried in solution or in suspension through streams and rivers into the oceans.

A second and important characteristic of water is its capacity to store heat. Water has a very high heat capacity, hence it stores heat during the day (or summer) and releases it slowly to the cooler atmosphere at night (or during winter).

The capacity of the oceans to store heat moderates the climate of neighboring land areas. Land, with its lower heat capacity, warms up more rapidly than the oceans during the day and cools quicker at night. On local geographic scales, such differences in heating and cooling rates between land and ocean result in local wind patterns that give rise to warm offshore breezes at night, and cool onshore breezes during the day.

Without its extensive surface waters, the Earth would be intolerably hot during the day and frozen at night. The vast extent of the oceans, therefore, acts as an enormous heat storage engine, slowly absorbing and releasing heat. Through the ocean basin circulation systems, heat is absorbed from the Sun's radiation in one area and transported to other areas, often at considerable distances, before being released to the Earth's atmosphere.

Density and pressure

The mobility of water is clearly an important physical characteristic. It has numerous consequences in terms of the transportation of heat and dissolved materials. Changes in density (as a result of expansion and contraction) combined with changes in salinity (due to inputs of freshwater from either land or from the atmosphere) result in the thermohaline circulation (movement driven by density differences) of the oceans. This circulation results, over time, in the movement of both surface waters from the tropics to high latitudes and deep-water flows of cold water from polar regions to lower latitudes. Such current systems play an important role in regulating the climate of the Earth and in determining the distribution of biological productivity in the ocean basins.

Several other physical properties of water have practical and important consequences for organisms both in the sea

▼ ***The clear blue*** *Mediterranean waters off the island of Ustica, southern Italy, provide excellent conditions for snorkeling. The clarity of the Mediterranean Sea is due to its low concentration of dissolved and suspended materials. A Secchi disk is often used to measure the clarity of water.*

▶ ***When light hits*** *the surface of the sea, between 3 and 30 percent reflects back, depending on the angle of the sunlight to the surface. The ocean's characteristic blue color results from absorption and scattering of other wavelengths of light – the red end of the spectrum is the first to be lost. In areas where suspended sediment is high, scattering of light occurs and the ocean may appear greenish or brown, depending on the particles present. At depths below 3300 ft (1000 m) virtually no light penetrates at all.*

and on land. Although water in general is not considered compressible, the pressure on the water in the deep-ocean basins is such that it is thought that the water column is compressed by around 100 feet (30 meters) under its own weight. If water were incompressible then sea levels would be higher than today and areas of productive, low-lying land would form part of the seafloor.

Water and light

The optical properties of seawater are also of vital importance to life in the oceans. The primary producers (phytoplankton) – organisms that convert sunlight into chemical energy through the process of photosynthesis – are dependent on sunlight and nutrients. Ultimately, all other marine organisms are in turn dependent on phytoplankton for their sources of food. Seawater, with its high concentrations of dissolved and suspended matter, absorbs light. Where dissolved and suspended materials are at low concentration, such as in the Mediterranean, the clarity of the water is high so that sunlight penetrates to greater depth and a swimmer on the surface can see to greater depths. Attenuation of sunlight in the oceans occurs to such an extent that below about 650 feet (200 meters) no light penetrates at all. Different wavelengths of light penetrate to different depths and the consequence is that most primary production is confined to a few tens of meters below the surface.

The depth to which only 1 percent of sunlight penetrates is called the euphotic zone. The photic zone, the depth at which photosynthesis occurs, is limited by the penetration of wavelengths of light in the range between 400 and 700 nanometers. In general, red light penetrates least and blue-green light the most, although this situation may be reversed where dense phytoplankton blooms occur and high-chlorophyll concentrations absorb the light in the blue-green end of the spectrum. Such differences in light absorption can be used by ocean color-scanners, mounted on satellites, to measure primary production on the surface of the ocean.

Chemically, seawater is an unusually pure substance, being almost 95 percent water. The remarkable chemical properties of seawater are a reflection, at least in part, of the substances dissolved in it. The dissolved elements are ultimately derived from land through erosion of the land surface, which results in material entering the ocean in suspension and solution.

Chemical composition

In addition to oxygen and hydrogen, the basic components of water – the two most abundant elements in seawater – are sodium and chlorine, which together produce salt, with concentrations around 1.05 and 1.9 grams per liter, respectively.

Salt concentration of seawater is expressed in parts per thousand, and for open ocean seawater, the concentration is around 35 parts per thousand. However, the salinity of the ocean varies regionally according to the local inputs of freshwater as rain and run-off from the neighboring land, and depending on the rate of evaporation from the ocean surface. In areas of high freshwater input, surface salinity may be as low as 32 parts per thousand, whereas landlocked seas may have much higher salinity – between 40 and 41 parts per thousand, as in the Red Sea.

Of the other 80 elements that occur in seawater, a number are concentrated by marine organisms – a process known as bioaccumulation. For example, vanadium has a concentration in seawater of less than one millionth of that of sodium, yet some filter-feeding animals have been found to accumulate concentrations 100,000 times greater than that of the surrounding seawater. Other bioaccumulators include oysters, which absorb zinc; lobsters, which concentrate copper; and several shellfish, which concentrate mercury. Bioaccumulation of mercury by shellfish led to the poisoning of hundreds of people in Minamata, Japan, in the 1950s and 1960s.

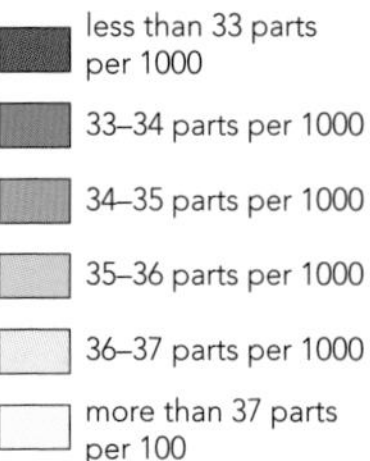

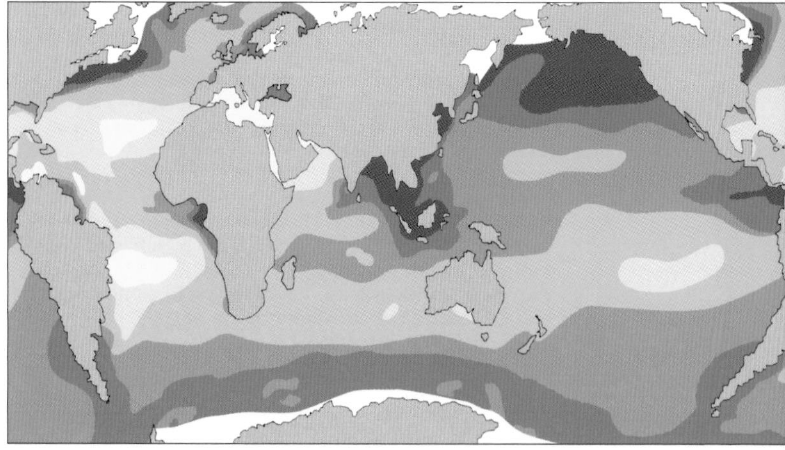

▲ ***A spectacular bloom*** *of phytoplankton off the coast of Norway, 2003. The influx of nutrients from the discharge of rivers is one reason why such blooms are common in coastal areas. Cold water from deep in the ocean may also well up to the surface and displace surface waters that have become depleted of nutrients.*

◀ ***The map*** *shows the average surface salinity of the oceans for the month of February. The areas of highest salinity are found in semienclosed seas such as the Caribbean and the Red Sea where evaporation of surface water exceeds inputs of freshwater from run-off and rainfall. The areas of lowest salinity are found in areas where large rivers, such as the Yangtze, Mississippi and Amazon, discharge into the ocean.*

Ocean nutrients

The nutrients that are essential for the growth of phytoplankton occur in solution in seawater, including nitrogen, phosphorus and silica. As these substances are removed from solution in the surface waters of the oceans by the phytoplankton, primary production declines. Only when these nutrients are released back into the water can a new cycle of production begin. Human disposal of sewage and run-off of agricultural fertilizers may alter the availability of nutrients in coastal waters, leading to bursts of primary productivity in the form of algal blooms.

Once the phytoplankton die, they start to sink and decompose. Those that sink below the level of light penetration (the euphotic zone) carry with them the nutrients they acquired during their lifetime. Together with other organic matter, these are decomposed at depth by bacteria, and the nutrients are released once more into solution. This process of decomposition results in a decline in dissolved oxygen concentration. In areas where water forms layers in semienclosed bays, the bottom water may become anoxic (without oxygen) and fish die.

Until the nutrients are brought to the surface, they cannot be used by photosynthetic organisms, the distribution of which is limited by the depth of light penetration. In coastal areas, mixing of the surface and bottom water may only occur on a seasonal basis under the influence of changes in temperature and wind direction. Some of the most productive areas of the open ocean are those where cold, nutrient-rich water is brought to the surface in the zones of upwelling.

Seawater elements

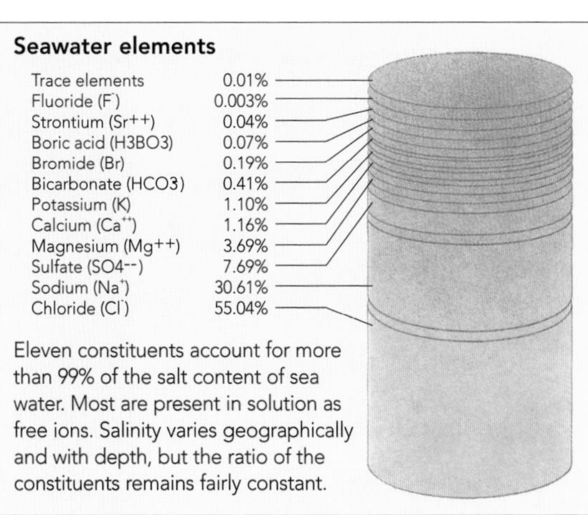

Eleven constituents account for more than 99% of the salt content of sea water. Most are present in solution as free ions. Salinity varies geographically and with depth, but the ratio of the constituents remains fairly constant.

The polar ice sheets, which cover about 12 percent of the surface of the oceans, have important effects on the world's climate. Ice acts as an insulating layer on the ocean surface, preventing further loss of heat from deeper water to the colder atmosphere above.

During the winter months, sea ice extends over an area of 6 million square miles (15 million square kilometers) of the Arctic Ocean, shrinking to about half that during the summer. During winter, the Antarctic ice cover extends to cover an area of some 8 million square miles (20 million square kilometers), which in summer shrinks to 1.2 million square miles (3 million square kilometers) of ice-covered sea. This ice cover restricts heat exchange between the ocean and the atmosphere, and water directly below the ice is much more variable in temperature and density than is normal for ocean surface waters.

It has been calculated that if all the Antarctic ice sheet were to melt, then the global level of the sea would rise by as much as 200 feet (60 meters). When released from this enormous weight of ice, the entire landmass of Antarctica would rise by between 650 and 1000 feet (200 and 300 meters). It is important to realize that melting of sea ice does not affect global sea levels since the current weight of the sea ice floating on the sea surface is already affecting global sea level.

There is increasing evidence that interannual variations in ice cover may have profound effects both on the ocean and atmospheric circulation patterns around the poles. Ice cover affects the exchange of heat and moisture between the ocean and atmosphere, which in turn affects air movements and both wind and surface sea currents.

▼ ***Midnight sun*** *reflects off the ice sheets of the Weddell Sea, Antarctica. The sun does not set in Antarctica during summer and the ice sheets shrink considerably during these months.*

Sea ice

Apart from icebergs, which are calved from the glaciers at the edges of land-based ice sheets, all sea ice is formed as the sea surface freezes. Seawater freezes at 29°F (–1.9°C), during which the surface of the sea undergoes a series of recognizable changes. At first, small crystals of ice develop which become packed together into either a continuous skin or a series of pancakes with characteristically raised rims. The first crystals to appear are minute spheres, which develop into thin disks or platelets known as 'frazil' ice. Slightly later, when the crystals have multiplied, the sea takes on a soupy, matt appearance, known as 'grease' ice. If the sea is calm, then the ice crystals freeze to form a continuous semitransparent skin called nilas. But if the sea is slightly rough, then the nilas breaks up into individual plates, which have raised edges due to

The Arctic ice sheet

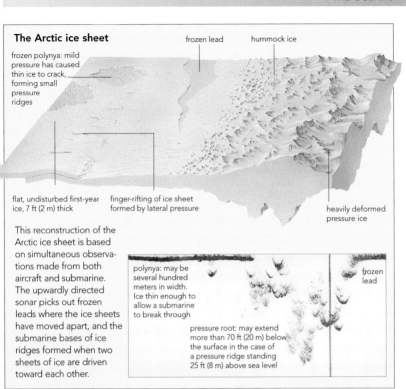

frozen polynya: mild pressure has caused thin ice to crack, forming small pressure ridges

frozen lead

hummock ice

flat, undisturbed first-year ice, 7 ft (2 m) thick

finger-rifting of ice sheet formed by lateral pressure

heavily deformed pressure ice

polynya: may be several hundred meters in width. Ice thin enough to allow a submarine to break through

pressure root: may extend more than 70 ft (20 m) below the surface in the case of a pressure ridge standing 25 ft (8 m) above sea level

frozen lead

This reconstruction of the Arctic ice sheet is based on simultaneous observations made from both aircraft and submarine. The upwardly directed sonar picks out frozen leads where the ice sheets have moved apart, and the submarine bases of ice ridges formed when two sheets of ice are driven toward each other.

their constant collisions with each other. Eventually, either the nilas thickens, or the pancakes freeze together to form a flat unbroken floe, which continues to increase in thickness, quickly at first, but more slowly as the thickening ice insulates the water beneath.

Salt-free ice

As the crystal lattice forms in the freezing water, salt is excluded and becomes concentrated into liquid brine cells, each less than 0.004 inches (0.10 millimeters) in diameter. Newly formed ice, therefore, contains less than 10 parts per thousand of salt compared with the 35 parts per thousand found in normal seawater. Eventually, the brine cells join together to form drainage channels through which the brine escapes into the sea.

Ice movement

As the ice coalesces and consolidates during the polar winter, the sea becomes covered with a layer of first-year ice, about 7 feet (2 meters) thick in the Arctic and 10 feet (3 meters) thick in the Antarctic. During the polar summer, the surface ice melts, forming surface pools. If the temperature is sufficiently high, the pools may extend through the thick ice sheet. If the floe survives the warmth of the summer months, it may become multiyear ice, which is extremely strong, multilayered ice that has sheets of refrozen meltwater incorporated into its structure. Such ice is blue in color and is common in the central Arctic, making it impassable to even the most powerful icebreakers.

An ice floe does not remain smooth for long. It is deformed by wind and current action, and may fracture to produce either long, open water leads, or polynya – wide pools of open water which quickly freeze. When winds subsequently converge, the new ice in a lead may become compressed and fractured and pushed up into pressure ridges. Pressure ridges may reach 32 feet (10 meters) in height and lie as deep as 150 feet (45 meters) below the ice sheet.

Wind stress is transmitted by the ice to the underlying water so that ice drift and current flow are identical over the long term. The Arctic has two major drift patterns: the Beaufort Gyre, in which ice floes and bergs may be trapped for up to 20 years, and the Transpolar Drift Stream, which carries ice from Siberia across the Pole and down the east coast of Greenland. In the Antarctic, winds blowing from west to east around the globe generate northeast-flowing currents that carry ice into the southern oceans.

▼ ***Antarctic icebergs*** *are generally larger than those of the Arctic, and are formed of pieces detached from the floating ice sheets fringing Antarctica, rather than being calved from the land ice sheets.*

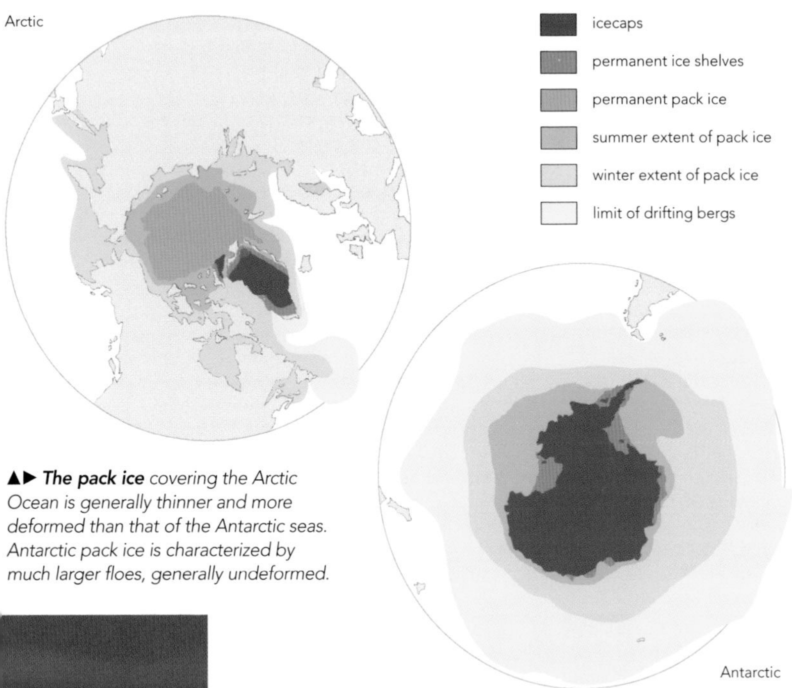

▲▶ ***The pack ice*** *covering the Arctic Ocean is generally thinner and more deformed than that of the Antarctic seas. Antarctic pack ice is characterized by much larger floes, generally undeformed.*

Floating giants

Icebergs, such as the one which sank the *Titanic*, are huge fragments of ice broken from the ends of glaciers or the edges of land-based ice sheets. A much greater danger to shipping than icebergs is superstructure icing, when the ship's center of gravity becomes higher, making the ship less stable and causing danger of capsizing.

In the Arctic, some 12,000 icebergs are calved annually from the glaciers. These drift with the current up the west coast of Greenland and return down the western coast of Baffin Bay, eventually passing out into the Atlantic. The survival of such icebergs is highly variable – in 1958 only one reached the Atlantic, but in the following year some 693 were recorded. A typical new berg weighs around 1.5 million tons, stands 260 feet (80 meters) out of the water, and extends more than 1150 feet (350 meters) below the surface. By the time it reaches the Atlantic, it may have shrunk to a tenth of its previous size.

Antarctic bergs are calved from the floating ice shelves that fringe that continent. Known as 'tabular' icebergs, they tend to have sheer sides and a distinctive flattened top. They are much larger than Arctic bergs – often more than 50 miles (80 kilometers) long – and usually retain their flattened tabular form. Thousands may calve each year, drifting north as far as 40°S before melting away.

The ancient Greeks were among the first to note the relationship between the tides and the Moon's monthly orbit around the Earth. However, it was not until Newton presented his gravitational theory, nearly 2000 years later, that the explanation for this relationship was more fully understood.

The lunar cycle

On the side of the Earth closest to the Moon, the gravitational force of the moon is strongest, causing the surface of the ocean to bulge toward the Moon. On the opposite side, the force is weakest, resulting in a bulge away from the Moon caused by the Earth's rotation. These forces and the response of the ocean would result in twice daily, or semidiurnal, tides perfectly aligned with the Moon were other factors not involved.

Since the Earth is spinning on its own axis, the bulge of the tides is displaced and appears slightly ahead of the Moon's actual position. This displacement is a result of the frictional forces between the water mass and the Earth's surface, which slows the oceans' response to the gravitational pull of the Moon. The Sun also exerts a gravitational force on the surface of the ocean, although this is weaker, around two-fifths as strong as that of the Moon.

The relative positions of the Sun and Moon result in monthly cycles of tidal amplitude. When both the Sun and Moon are

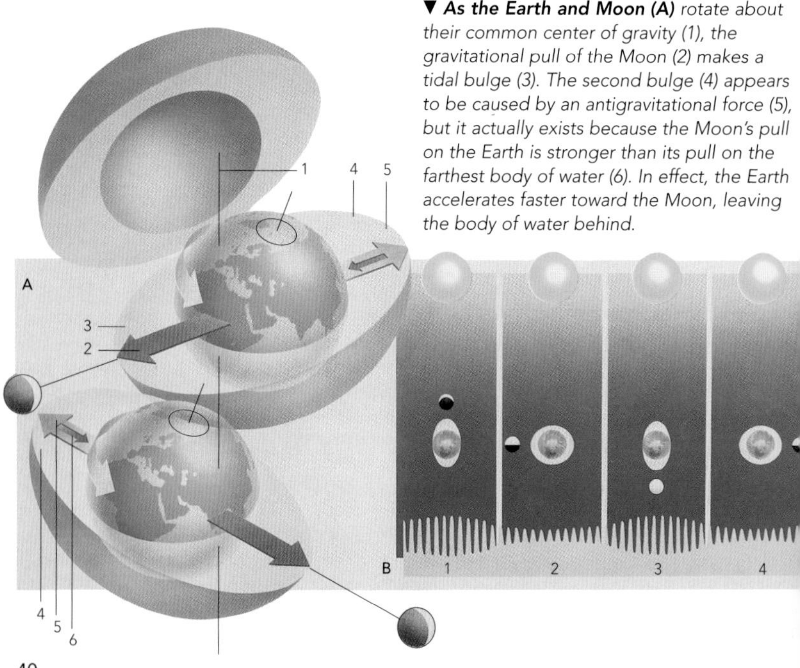

▼ ***As the Earth and Moon (A)*** *rotate about their common center of gravity (1), the gravitational pull of the Moon (2) makes a tidal bulge (3). The second bulge (4) appears to be caused by an antigravitational force (5), but it actually exists because the Moon's pull on the Earth is stronger than its pull on the farthest body of water (6). In effect, the Earth accelerates faster toward the Moon, leaving the body of water behind.*

▲ ***Moncton tidal bore*** *moves up the Peticodiac River, New Brunswick, Canada. The phenomena is caused by high tides in the Bay of Fundy, Nova Scotia.*

pulling together, the tidal range is greatest, resulting in high spring tides, but when the gravitational forces of the Sun and Moon are at right angles to one another, low neap tides result.

The Moon's orbital plane changes on a cycle of 18.6 years, reaching a maximum angle of 28.5° to the Earth's equator. At such times, the Moon changes its position from 28.5°N to 28.5°S of the equator every lunar month. When the angle between the Moon's orbital plane and the equatorial plane are greatest, the two daily tidal heights are different, with one very high tide and one smaller high tide.

Tidal variations

In addition, the oceans' water mass is not free to flow over the entire surface of the globe – being restricted in the different ocean basins, and with internal flow being channeled by the midocean ridges and other topographic features. The oceans are also acted on by the Coriolis forces generated by the Earth's own rotation. The result is that the tidal patterns differ in different oceans, reflecting the interaction of this wide variety of forces. The tidal rise and fall of the ocean surface is generally between 3 and 10 feet (1 and 3 meters), but may be much greater in some semienclosed bays, such as the Bay of Fundy in Nova Scotia where the tidal range reaches 42 feet (13 meters).

◀ ***Spring tides (B),*** *the highest tides, occur when the Sun, Moon and Earth are in a straight line (1 and 3). When the Sun and Moon are at right angles to one another (2 and 4), their gravitational forces to some extent cancel each other out, producing neap tides.*

Most coastal areas experience semidiurnal tides, and diurnal tides are relatively rare due to the shape of the coastline. Mixed tides are an amalgam of these two tidal types and are characterized by two tides each day, one of which is much stronger than the other. The Atlantic and Indian Oceans generally display semidiurnal tides which are of similar magnitude, while the Pacific Ocean displays a more mixed tidal pattern in which one tide each day is much larger than the other.

CURRENTS AND GYRES

The major patterns of water movement in the ocean basins determine the overall direction of current flow. Near a coastline, the pattern of coastal currents is more obviously influenced by tidal forces, the direction of riverine flows and the shape of the coastline itself. Such local patterns of water movement dominate inshore waters in the continental shelves, but the waters of the open ocean move in response to processes occurring over much wider geographic scales.

The great current systems of the open ocean are driven by atmospheric circulation. Once away from the influence of neighboring land, large-scale patterns of water movement dominate the ocean basins. These patterns include the great gyres, or circular movements of the ocean north and south of the equator, and major surface currents, such as the circumpolar current of the Antarctic and the equatorial currents, which cross the Pacific and Atlantic Ocean basins.

Convection currents

Convection currents move surface waters from the warmer equatorial regions toward the poles where the water becomes colder and denser, sinking to great depths before moving back toward the equator. Vertical movements of water in the oceans are influenced by density, which in turn is affected by salinity and temperature. As water evaporates from the sea surface in the tropics, it becomes more saline and thus denser. At the same time, increasing temperature results in expansion and a decrease in density.

In the tropical Atlantic, water loss from evaporation is greater than the gains from rivers and rainfall, and the sea surface becomes warm and saline, flowing northward as the Gulf Stream. As this saline water moves north, it loses heat to the atmosphere, and by the time it reaches Greenland, it is

January temperature and ocean currents
(Northern Hemisphere – winter)

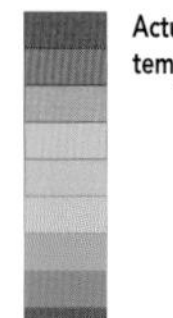

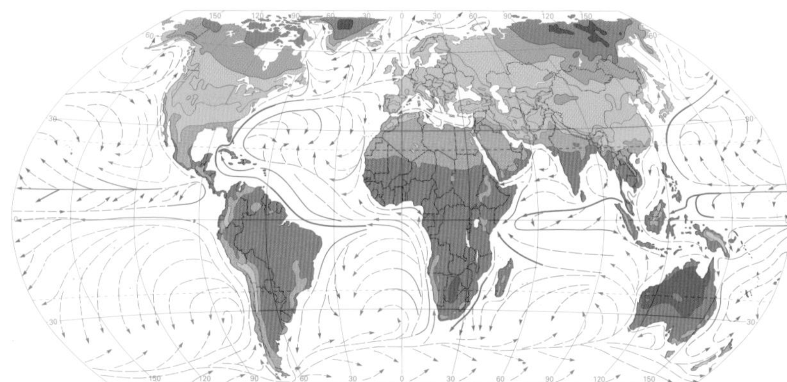

▲ ***Ocean currents*** *are primarily dependent on wind patterns. These in turn depend on seasonal changes in temperature. Depending on the annual movement of wind belts, some currents on or near the equator may reverse their direction in the course of a year.*

near freezing. The surface waters are now more dense than the underlying water, convection sets in, and the surface water sinks to flow southward along the eastern margin of the Atlantic. Deep, cold water masses also form in the seas near the Antarctic, flowing northward along the western margins of the southern continents.

Current patterns

The wind systems in each hemisphere are dominated by the trade winds (in the vicinity of the equator), which drive the surface ocean waters to the west and by the westerly winds at higher latitudes, which return water to the eastern margins of the ocean basins. The combined effect of west-moving waters at low latitude and east-moving waters at higher latitudes results in the circular current patterns, or gyres. A gyre is a basin-scale, closed-current system which consists of a strong western boundary current carrying the water poleward and a less strongly defined return flow to the east.

In the North Atlantic Gyre, the North Equatorial Current flows from Africa toward Latin America, becoming the northerly flowing Gulf Stream, which flows along the North American eastern seaboard before recrossing the Atlantic as the North Atlantic Current, then flowing south as the Canaries Current which rejoins the North Equatorial Current to complete the cycle.

The speed of the ocean surface currents is generally around 6 miles (10 kilometers) a day, but the western boundary currents, such as the Gulf Stream of the North Atlantic and the Kuroshio Current in the North Pacific, may achieve speeds of up to 60–100 miles (95–160 kilometers) a day. In the southern hemisphere, the currents are generally weaker than in the north, and the more open, southern ocean system is dominated by the Antarctic Circumpolar Current.

July temperature and ocean currents
(Northern Hemisphere–summer)

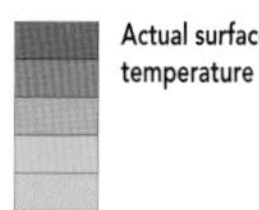

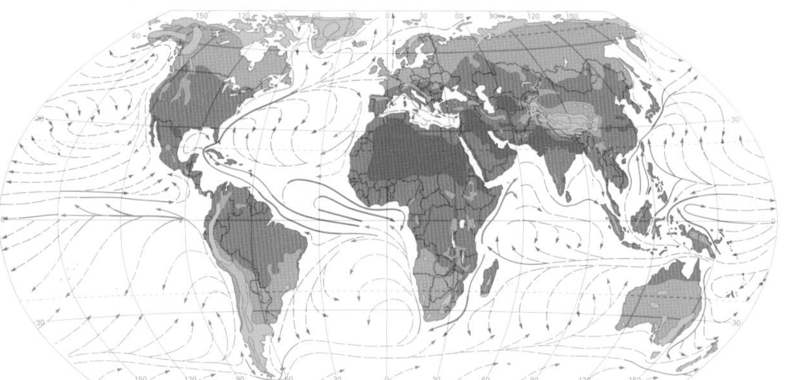

▲ ***The 38 major named currents*** *make up five current gyres – the North Atlantic Gyre, the South Atlantic Gyre, the South Indian Gyre, the North Pacific Gyre, and the South Pacific Gyre. Only the main currents are named on the above maps.*

The Ekman spiral

Although persistent winds drive the ocean current systems, other forces also act on the moving water mass to change its direction and speed of movement. The so-called Coriolis force results from the rotational movement of the Earth which tends to deflect any moving particle, whether water or air, relative to the surface of the Earth. If a water mass moves northward in the Northern Hemisphere, the Earth's rotational influence results in it moving clockwise, or to the right; while a southerly moving mass in the Southern Hemisphere moves to the left, or anticlockwise. This contributes to the flow of water in a gyre.

The spiral movement of water in the gyres results in water being piled up toward the center. The level of the water in the Sargasso Sea, for example, is about 3 feet (1 meter) higher than in adjacent coastal regions. The outward pressure of this dome of water balances the inward pressure created by the rotational forces of the Earth's spin. Changes in atmospheric pressure can also influence the level of the sea, and stable areas of high or low atmospheric pressure can cause rotational movements of the surface water around such centers.

When winds blow over the surface of the ocean they impart motion to the surface of the water through friction, thus the surface layer of water tends to move in the same direction as the prevailing wind. As one passes deeper into the water mass, however, the influence of the Coriolis force becomes more apparent and the direction of water movement becomes progressively deflected away from the direction of surface movement. This so-called Ekman spiral

▼ ***Two large ocean*** *circulation features, called eddies, at the northernmost edge of the ice pack in the Weddell Sea, off Antarctica. The eddy processes in this region play an important role in the circulation of the global ocean and the transportation of heat toward the pole. This image was produced by a spaceborne radar.*

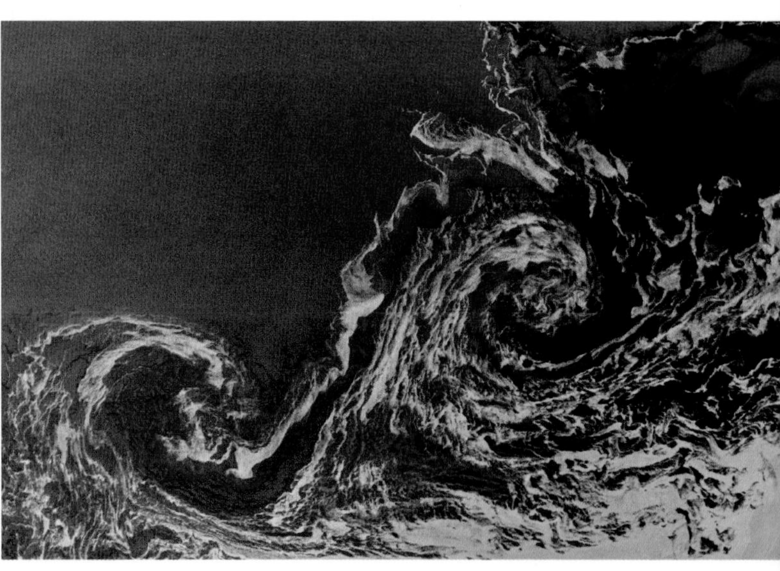

The Ekman Spiral

In 1893 Norwegian explorer Fridtjof Nansen proposed that the rightward drift in the northern hemisphere was due to the Coriolis force. In 1905 V. Walfrid Ekman mathematically proved this theory. The wind's frictional force tries to drag surface water after it. However, the water is immediately deflected by the Coriolis force. The surface layer in turn drags the layer beneath it, which again is deflected. As movement is transmitted downward, the deflections form an Ekman spiral, whereby the deepest water may be moving at 180° to the surface flow. These forces result in the surface current flowing at an angle to the wind.

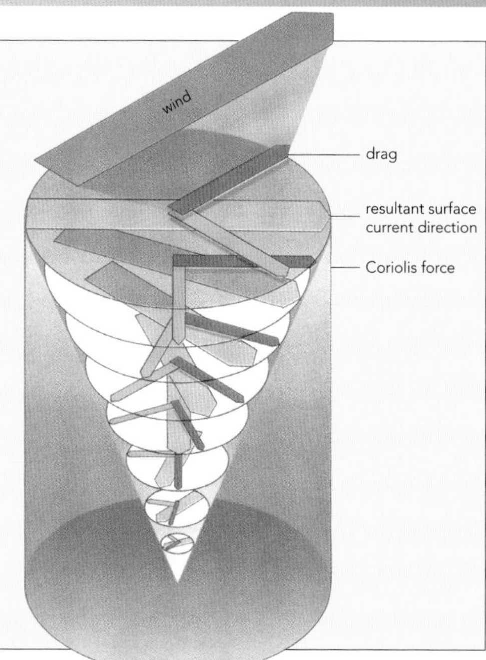

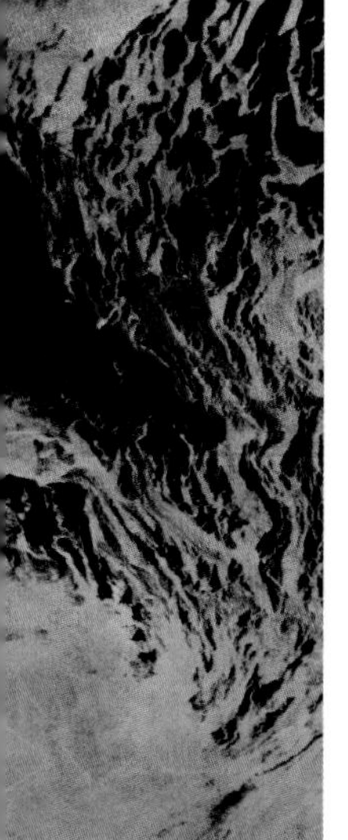

involves the upper 300 feet (90 meters) of water in the ocean surface. While the overall direction of movement of the entire water mass is at right angles to the direction of the prevailing wind, the surface current flows at an angle to the wind. At depth the water may be moving in the opposite direction to the surface currents.

Mesoscale eddies and 'rings'

As a consequence of the Ekman transportation of whole water masses, the trade winds and westerlies actually pile up water in the center of the gyre and result in the gyres being squeezed toward the western margins of the ocean basins. This contributes to the faster speeds of the Gulf Stream and other western boundary currents, which are generally fast flowing and stable; however, as they move to higher latitudes and begin to be deflected away from the coast, they lose their stability and begin to meander. These meanders increase in amplitude and may ultimately be cut off from the main flow as 'Gulf Stream Rings'.

The existence of mesoscale eddies and rings has only been known for a relatively short period of time and, with the advent of satellites, the improvements in remote sensing of the sea surface have demonstrated their wide distribution in association with strong ocean currents. The importance of these features in terms of ocean energy transfer and in the patchiness of ocean productivity is only now becoming better understood.

Waves are a familiar sight. Their awesome power releases a considerable destructive force as they break against a shoreline. Waves are of several different types. They differ in their length – the distance separating wave crests – and their periodicity – the time which separates successive wave crests. Waves are formed when wind energy from the atmosphere is transferred to the surface waters of the ocean, and waves break when they reach shallow depths.

Tidal forces acting on the ocean water masses produce complex standing and rotating wave systems characteristic of each ocean basin. Nodal points occur at the centers of these wave systems and represent the points of least disturbance of the ocean water mass where changes in the vertical height of the sea surface are generally around zero.

Capillary and wind waves

At one end of the scale are ripples, or capillary waves. They have a periodicity of less than 1 second and wavelengths of less than 0.69 inches (1.74 centimeters). At the other end of the scale are the tides, with a periodicity of 12 or 24 hours.

Capillary waves are set up by wind blowing gently across the water surface. Such waves are characterized by rounded crests and pointed troughs. Wave shape is determined by surface tension, and gravity only becomes important in waveform when the wavelength exceeds 0.69 inches (1.74 centimeters). As wavelength increases, the waves become more pointed and the troughs more rounded. Waves grow in height as increasing wind pressure acts against the waves' windward side. As the waves grow, their crests become steeper and the wind reinforces the wave shape by pressing against the windward side and eddying over the crest thus reducing pressure on the leeward side of the crest. Once the crests reach 120°, they become unstable and break, producing breaking waves characteristic of strong wind conditions.

▼ ***Satellite image*** *of Hurricane Floyd on September 14, 1999. Floyd was a large and intense Cape Verdean hurricane that pounded the central and northern islands of the Bahamas, seriously threatened Florida, struck the coast of North Carolina, and then moved up the east coast of the United States into New England. It neared the threshold of Category 5 intensity on the Saffir-Simpson Hurricane Scale as it approached the Bahamas, and created a flood disaster of immense proportions in the eastern United States, particularly in North Carolina, where it killed 52 people.*

Storm waves

Waves move across the surface of the ocean in much the same way as a sailing ship, with the wind behind the wave pushing it across the ocean surface. The energy transferred from the wind to the surface of the water can itself be transferred from one wave to another. During storms, waves receive so much energy that they are whipped into waves of different periods, lengths and directions. The resulting chaotic pattern is termed a 'wind sea'.

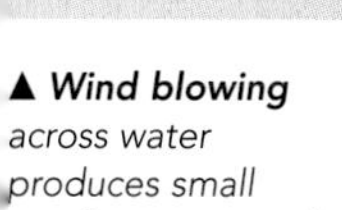

▲ ***Wind blowing*** *across water produces small capillary waves with round crests and pointed troughs. This shape is determined by surface tension.*

▲ ***Once wavelength*** *exceeds 0.69 in (1.74 cm), gravity takes over from surface tension as the dominant force on wave form. The crests become more pointed; the troughs rounded. The wind reinforces the wave shape by pressing down on the windward side and eddying over the crest to reduce pressure on the leeward side.*

▲ ***As the waves*** *grow, their crests steepen until they reach an angle of 120°, at which point they become unstable and break, producing whitecaps.*

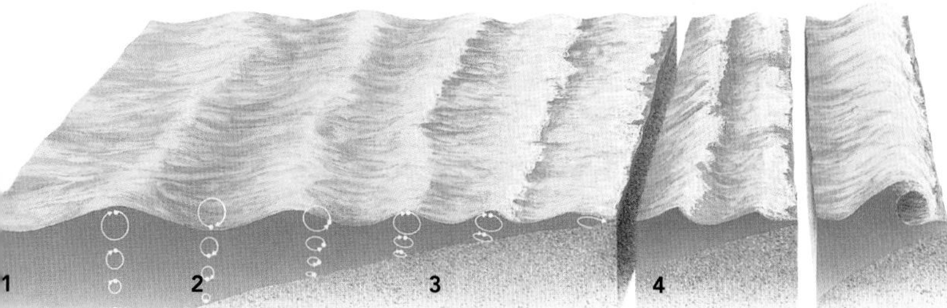

1 ***Waves in the*** *process of generation are known as 'forced waves'. As they move from the source area, they travel as long-crested swell waves.*

2 ***The orbital movement*** *of surface water particles has a diameter equal to the wave height, but reduces to zero at a depth equal to half the wavelength.*

3 ***In shoaling water,*** *waves heighten and steepen and wavelength shortens. Particle movements become more eliptical until they can no longer hold their orbits and the wave breaks.*

4 ***Breakers*** *on a gently sloping shore may surge up the beach (left) or break gradually as water spills down the wave front. On a steep shore (right), waves overturn as plunging breakers.*

Wave height, the vertical distance between the trough and crest of a wave, is determined by the fetch (distance over which the wind blows), the duration of the wind, and wind speed. In the Pacific Ocean, which has the greatest fetch of any ocean, wind waves reach their greatest height. At the heart of intense storms, waves may reach between 40 and 50 feet (12 and 15 meters), although the largest recorded wave was 115 feet (35 meters) high. A well-documented case of such a swell was recorded by the captain of the *Ramapo* – a 146 meter (478 foot) -long tanker outbound from Manila to San Diego – whose ship was caught in a storm during which the wind blew at 30 to 50 knots for several days. At the height of the storm, the tanker was passed by a wave 112 feet (34 metres) high that was traveling at an estimated 55 knots.

▲ ***In 1998 a tsunami*** *in Papua New Guinea destroyed the two Arop villages that stood on the sand spit in this photograph. The wave removed all trace of the several hundred houses that stood on the spit, except for the remains of a septic tank seen in the foreground of this image. Many of the 2100 victims drowned in the Sissano Lagoon behind the spit.*

High waves occur most frequently when storm waves drive up against the continental shelves. As waves move away from a storm center, they change shape, becoming less ragged and taking the form of long smooth swells. Such waves can travel very long distances and one such swell, caused by a storm in the Indian Ocean, is known to have traveled 12,400 miles (20,000 kilometers) to Alaska. In 1987, long-distance swells from a storm center in the Southern Ocean caused extensive flooding and damage in the Maldives.

Breakers and surf

As waves approach the shore, their height becomes greater and their wavelength shorter, eventually breaking as surf on the shore. The breaking of the wave results from the orbital movement of the water particles inside the wave and the friction caused by the bottom, which slows the movement of the bottom water relative to the surface.

Inside each wave, water moves in an orbital way with particles at the surface moving in a circle of diameter equal to the wave height. At a depth of half the wavelength, no orbital motion is present in the water mass. As the waves approach the shore and the depth decreases to less than half the wavelength, the particle movements become more elliptical until they can no longer maintain their orbit, and the wave breaks.

Where the fetch is greater than 800 miles (1300 kilometers) and the wind blows constantly for several days in the same direction, long distance swells are established which break as surf on the shoreline. Since the world's oceans extend for thousands of miles between the coasts of major landmasses, open ocean swells of great height can be formed, breaking as surf on the shore. Surf on oceanic islands, such as Hawaii's

north shore is generated by winter storms off Alaska, while summer surf on California's beaches originates from storm centers as far away as the Tasman Sea, and Puerto Rico's western beaches receive giant surf from winds blowing across the Caribbean all the way from the Texas coast. Since Atlantic storms form close inshore, the waves do not have sufficient distance to develop into the large swells that break as surf. On the other side of the Atlantic, off western France and the southern tip of England, the Atlantic swells do not have sufficient fetch and break as surf.

Tsunami

The so-called tidal waves, properly termed tsunami, are not associated with tides but with submarine earthquakes or volcanic eruptions. A sudden shift in the ocean floor caused by vertical movements along a fault line may create a wave at the surface of the ocean. These waves can travel at up to 450 miles (720 kilometers) per hour. The amplitude of such waves is usually quite low, and much of the wave energy is expended at the continental slope. However, in some areas, where the bottom topography focuses the wave energy, devastating effects on land can occur.

Although tsunami can occur in the Atlantic, they are more frequent in the Pacific Ocean, where they are generated around the Pacific Rim, in association with the active subduction of the ocean-plate margins. As a consequence, systems have been established to provide early warning of the passage of such waves following earthquake activity.

▲ ***A tsunami*** *occurs when a submarine earthquake causes a sudden shift in the seafloor along a fault line. The upheaval creates a bulge that breaks down into a series of waves traveling at speeds of up to 450 mph (750 km/h).*

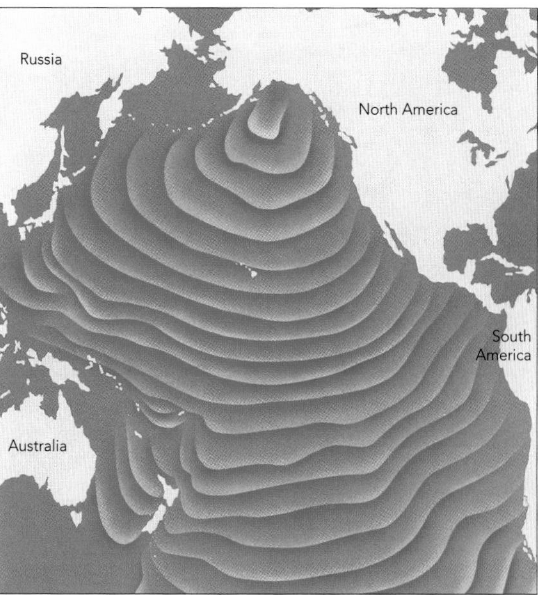

► ***The map*** *shows the hourly procession of the 1946 tsunami that originated in the Aleutian Islands.*

The oceans have been described as the flywheel of the climate system. They store energy when it is in abundant supply during the day, or summer, and release it during the night, or winter. Unlike a flywheel, however, the oceans play a much more active role in the global climate system. Through their constant motion, the oceans transport considerable quantities of energy across the surface of the Earth. The heat storage capacity of the oceans and the length of time involved in ocean circulation, play a major role in determining regional and global climates.

Temperature variations

When the ocean cools, it responds through the process of convection, resulting in heat being brought to the surface. The overall cooling is, therefore, spread over a considerable depth, and the overall fall in temperature of the surface is quite small. As a result, the surface of the world's ocean varies in temperature over a much smaller range than does the land surface, from 28° to 86°F (–2° to 30°C). At any one location, however, the variation in temperature is even smaller, less than 2°F (1°C) during the course of a day and around 18°F (10°C) over a period of one year. In contrast, the range of temperature over a continental landmass may be as great as 180°F (100°C) from place to place and as much as 144°F (80°C) in one site over the course of a year.

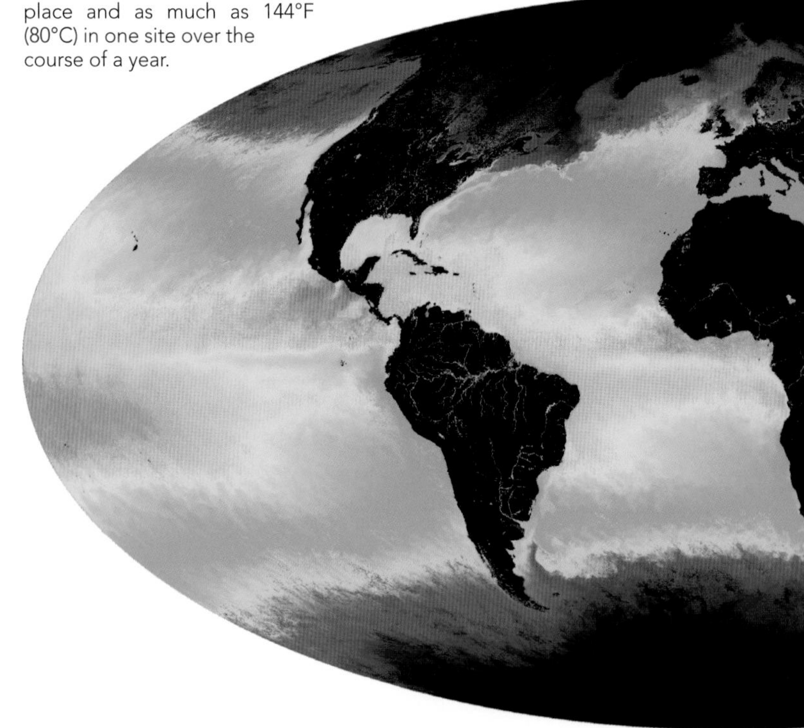

The thermal inertia of the oceans and their slow response results in delays in the seasonal cycles of ocean areas compared with land. This difference in timing of response gives rise to both short- and long-term changes in atmospheric circulation, of which the great seasonal monsoon cycles are a well-known example.

Monsoon circulation

The monsoons of the Indian Ocean are a good illustration of the interaction between the ocean and the climate system. During the northern hemisphere winter, high atmospheric pressure areas form, centered on the Asian landmass and forming cold, dry air masses which find their way through southeast Asia to become the northeast monsoon over the northern Indian Ocean. The northeast monsoon blows from northeast to southwest and drives the surface circulation anticlockwise in the Bay of Bengal and Arabian Sea during the months of December to February.

During the northern hemisphere summer, the desert areas of Africa, Arabia, Pakistan, China and India become very warm, creating zones of low pressure which attract moist winds from the southern parts of the Indian Ocean. These southeast trade winds bring high rainfall over India and southeast Asian countries and, after crossing the equator, become the southwest monsoon. Ocean circulation is reversed during the monsoon, with surface circulation flowing in a clockwise direction.

◀ ***Sea surface temperature*** *This false-color image shows a one-month composite for May 2001. Red and yellow show warmer temperatures, green is an intermediate value, while blue and then purple are progressively colder values. The cold water currents move from Antarctica northward along South America's west coast. These cold, deep waters upwell along an equatorial swathe around and to the west of the Galapagos Islands. Note the warm currents of the Gulf Stream moving up the east coast of the US, carrying Caribbean warmth toward Newfoundland and across the Atlantic toward Western Europe.*

The ocean heat engine

Since ocean waters move both vertically and horizontally over the Earth's surface, they redistribute considerable quantities of heat around the globe. The movement of warm, tropical waters to both north and south polar regions represents a source of heat transfer that is as large as that of the atmosphere and which has dramatic consequences for climates on adjacent landmasses. The North Atlantic Gyre represents one such heat engine, and the influence of the Gulf Stream results in mild winters in Western Europe.

▲ ***The Charleston bump*** *(yellow) is a rocky ramp in the Gulf Stream that causes the Stream to change direction, forming a permanent meander.*

Mild winters

In the tropical Atlantic Ocean, solar heating results in the evaporation of water from the surface and so causes a rise in salinity and hence in density. Some of this water flows north, passing between the coasts of Iceland and Britain. Here it gives up heat to the atmosphere, which, because the winds in this area are predominantly from the west, carries warm air across Western Europe. This flow of heat results in the mild winters characteristic of Western Europe and distinguish it from other continental areas at similar latitude, which are much colder.

In addition, because the water gives up heat to the atmosphere, the surface water temperature drops close to freezing point and the density of the water increases further. In the Greenland Sea, for example, the combination of low temperature and high salinity ensures the surface waters are denser than the underlying water. As a consequence they sink, occasionally to the bottom, where the cold, high-salinity water mixes with, and slides under, the bottom water, spreading on the ocean floor and flowing south.

This pattern of thermohaline circulation in the North Atlantic, which results in the mild climate of Western Europe, may not have occurred at the end of the last ice age. The rapid melting of the Laurentian ice sheet would have released considerable quantities of freshwater into the surface waters of the North Atlantic. Under such conditions, the North Atlantic would have been covered by a layer of less saline water, which would have frozen in some areas. This frozen water would have prevented the transportation of heat from the equatorial Atlantic and therefore resulted in severe winter climates in Western Europe.

Large quantities of cold, deep water are also formed in the southern hemisphere around the Antarctic landmass. Deep water from the north and south polar regions spreads out through the world's oceans, becoming mixed with the warmer water above as fresh supplies of colder water push

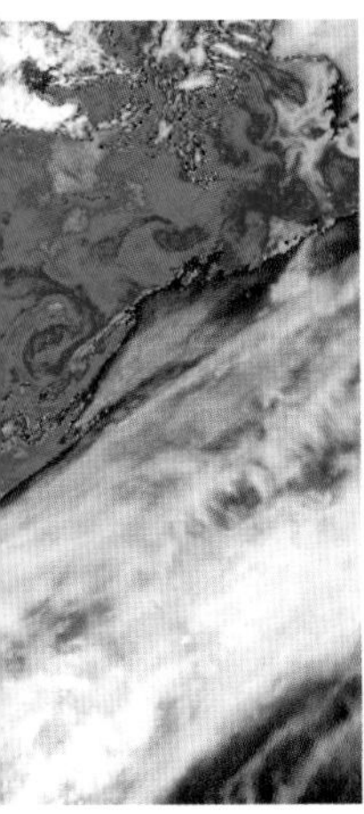

beneath it. Over the centuries, these cold-water masses gradually rise, close enough to the surface to become involved in surface water mixing. Shallow circulation of water, which does not reach very high latitudes, but nevertheless passes poleward, cools and sinks as a consequence of convection and may circulate at intermediate depths on timescales of a few years or decades.

Marine productivity

The circulation of deep-ocean water and the convection currents which cause it to sink in polar regions, are also vitally important for marine primary productivity and associated food webs. In areas of upwelling, where cold, nutrient-rich water is brought to the surface, high biological production occurs. Short-term changes in ocean circulation patterns can impact on marine productivity in such areas.

An example of seasonal productivity that occurs on an annual basis is found off the coasts of the Arabian Peninsula and Somalia. Off these coasts cold, nutrient-rich, water is brought to the surface from depths of 330 to 660 feet (100 to 200 meters) during the Southwest Monsoon season, and marine productivity is very high. Since the currents in the northern Indian Ocean are reversed during the southwest and northeast nonsoon seasons, upwelling does not occur during the northeast monsoon, hence productivity in this region is highly seasonal.

▼ ***Map of thermohaline circulation*** *In downwelling areas, heat transfers to the air. In upwelling areas, heat transfers from air to ocean.*

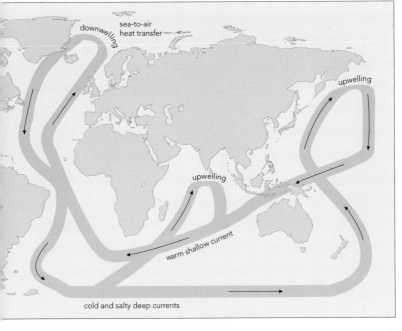

downwelling
sea-to-air heat transfer
upwelling
upwelling
warm shallow current
cold and salty deep currents

Almost every year, the American Gulf region or eastern coastline is hit by a hurricane accompanied by torrential rains and often devastating winds. Such storms are generated far out to sea off the coast of Africa or in the Caribbean region, where the temperature of the ocean surface waters is high. Hurricanes are perhaps the most awesome of the many manifestations of air-ocean interaction.

The vast ocean surface, covering some two-thirds of the Earth, acts as a giant thermostat. It absorbs energy in the form of heat during the summer and releases it over winter. The energy released by the oceans heats the atmosphere above, causing turbulent air movements. In turn, the movements of the atmosphere affect the surface of the oceans, establishing wind-driven waves and currents. Such phenomena may be of short or long duration depending upon the strength and persistence of the wind.

The higher density of water, compared with that of air, results in slower rates of change in movement of the surface waters of the oceans compared to the speed of response of the atmosphere to the heat released from the oceans.

Coastal breezes

The surface waters of the oceans are, however, intimately linked with the atmosphere above. This is perhaps nowhere more obvious than in short-term weather events such as the daily pattern of onshore, offshore breezes that occur in many coastal areas all over the world. Because the land is heated

▼ ***Time-lapse*** *satellite images of Hurricane Andrew, August 1992. Andrew is one of only three Category 5 hurricanes known to have struck the USA. Category 5 storms have winds of more than 155 mph (250km/h) and storm surges generally more than 18 feet (5.5 meters) above sea level. On August 24, Andrew cut a destructive swathe through southern Florida and entered the Gulf of Mexico. On August 26, it made landfall, 100 miles (160 km) southwest of New Orleans, Louisiana. It was downgraded the next day to a tropical depression, northeast of Jackson, Mississippi. The eye of the hurricane is visible in each image.*

Weather systems

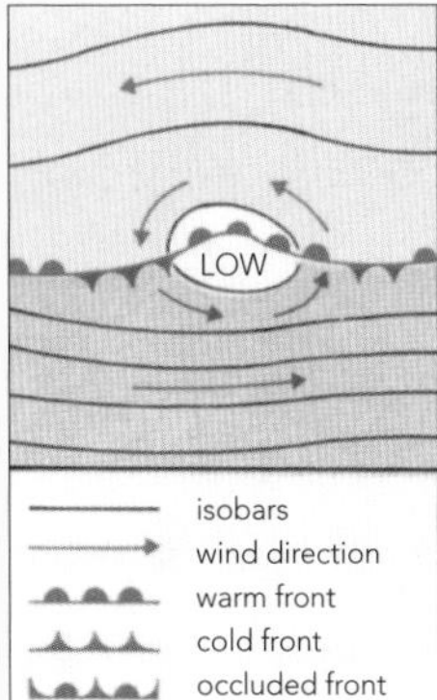

▲ ***Cyclones result** in much of the mixed rainy weather of temperate latitudes. They originate as waves in the polar front between polar air and maritime air.*

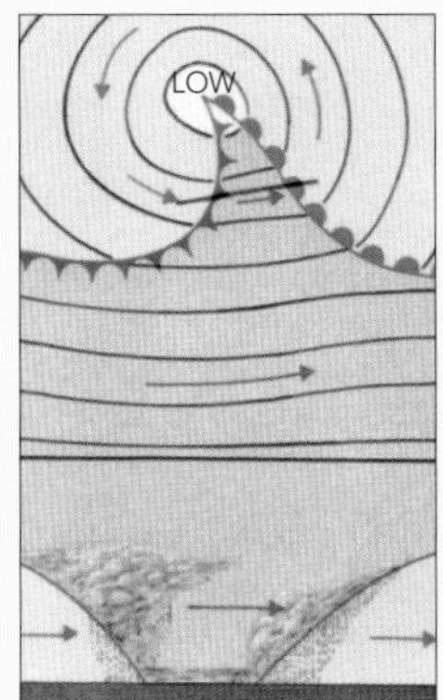

▲ ***These weather** systems are closely tied to longer waves in the upper atmosphere which affect the change of heat, momentum and moisture between the air masses.*

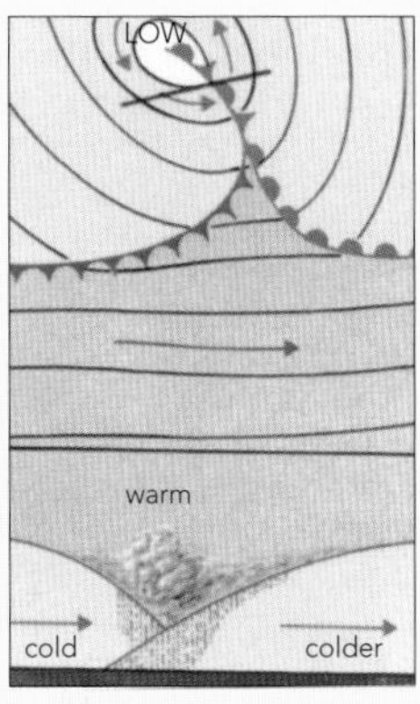

▲ ***The depression** tightens as it develops and occludes as the trailing mass of cold air catches and either undercuts or, as shown above, overrides the warm front.*

more rapidly than the sea during the day, the air above the land rises, and is replaced by air flowing from the sea to the land. Cool onshore breezes thus help to moderate the heat of the day. At night, however, the direction of the wind reverses. As the land cools down to a temperature below that of the adjacent sea, gentle offshore breezes spring up.

Another feature of daily weather reflecting the interaction between air and ocean is the tropical rainstorms that occur on small oceanic islands. During the day, water evaporates from the sea surface leading to a buildup in the water vapor in the atmosphere above the island. As the air cools later in the day, thunderclouds form and rain falls, often in short-lived but torrential downpours.

In addition to the transfer of heat from ocean to atmosphere, water vapor also enters the atmosphere from the surface of the seas. Where wind blows over warm ocean water, it picks up moisture, and if it subsequently encounters colder ocean water, dense ocean fog forms as water droplets condense in the atmosphere. In the northwest Pacific, for example, during the summer months, southerly winds flow over the warm Kuroshio Current and, as they pass over the cold Oyashio Current, they give rise to areas of dense ocean fog.

Tropical storms

The formation of storms (and in particular hurricanes) represents another, often violent, interaction between the ocean surface and the air above. In tropical areas, where the surface of the sea exceeds 80°F (27°C), heat and water vapor are discharged into the atmosphere. As the air is warmed, it becomes less dense and rises rapidly, forming a spiral pattern. As the warm, moisture-laden air rises and expands, energy is released through the heat of condensation, and the speed of the wind increases. The center, or eye, of such a storm forms a calm region of low pressure, while the dense walls of cumulus-type clouds, which encircle the eye, make up the regions of highest wind speeds and extremely dense precipitation.

Hurricane Gilbert, which swept through the Gulf of Mexico in 1988, was 900 miles (1500 kilometers) in diameter, and although the storm only traveled at 12–16 miles (18–25 kilometers) per hour, wind speeds near the eye reached more than 200 miles (320 kilometers) per hour. Pressure in the eye was a record low at 885 millibars. Wind speeds in such tropical storms are inversely related to the air pressure at the center – the lower the pressure, the higher the wind speed.

Once a hurricane or typhoon passes over land, it begins to lose its power, since the source of energy from the warmer ocean is cut off and the increased friction from the land surface upsets the pattern of air circulation. Associated with

▼ ***A waterspout*** *occurs when a tornado passes over water, sucking up a column of water up to 300 ft (100 m) tall. They usually occur in tropical regions, associated with stormy weather.*

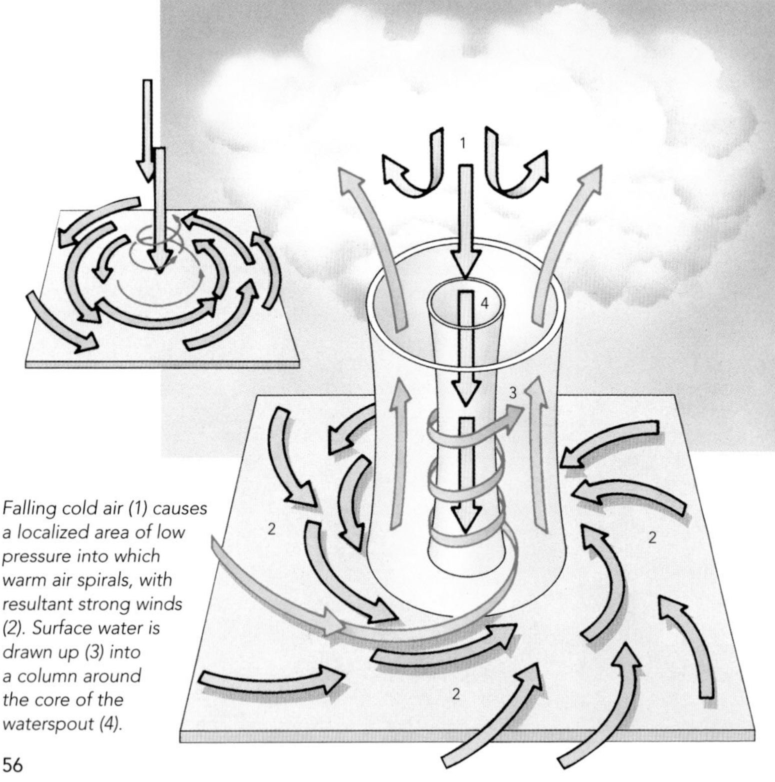

Falling cold air (1) causes a localized area of low pressure into which warm air spirals, with resultant strong winds (2). Surface water is drawn up (3) into a column around the core of the waterspout (4).

such storms are sea surges and torrential rain. Hurricane Gilbert was accompanied by a wave surge of 20 feet (6 meters) and rainfall of between 10 and 15 inches (250 and 380 millimeters), which fell in only a few hours.

Shaping the coasts

In terms of hours and days, atmospheric movements may result in considerable changes in the surface waters of the oceans. Where strong winds combine with high tides, the water level at the coast may be much higher than normal. This often results in penetration of salt waters into estuaries and coastal flooding.

Most coastal erosion, however, occurs over much longer periods of time as a consequence of storm waves generated during a particular season of the year by strong atmospheric disturbances. In many areas, beaches change their shape from season to season. This reflects seasonal differences in the predominant wind direction and consequent changes in the sites of deposition and erosion along the shore. Winds that are at a slight angle to the shoreline result in the lateral movement of the water mass (along the shorelines), which in turn causes drift of sediment materials in the direction of the current.

▲ ***A waterspout*** *or vortex photographed over the ocean. A boat is visible just in front of the column of water.*

Studying the air-ocean interface

An understanding of the mechanism by which air and ocean exchange heat and moisture is of fundamental importance to medium- and long-term weather forecasting, which in turn has enormous influence on agriculture and associated industries. Over the last few decades, considerable international time, effort and money has been invested in the investigation of weather and the climate system. In the 1970s, the Global Atmospheric Research Program (GARP) was established to examine these interactions on an ocean basin scale.

The GARP Atlantic Tropical Experiment (GATE) experiment in 1974 investigated (through a coordinated program of ship-borne, aircraft and satellite observations) air-sea interactions in the tropical regions of the Atlantic. In the 1980s the TOGA (Tropical Ocean Global Atmosphere) program was established to examine similar air-ocean interactions in the western Pacific. TOGA has, for example, contributed substantially to our understanding of the El Niño phenomenon. During El Niño years, a major change occurs in the ocean currents of the southern Pacific with consequences for climate, weather patterns and sea level conditions over wide areas of the tropics.

The importance of the air-sea interaction is nowhere more dramatically demonstrated than in the Southern Pacific Ocean. Periodic weakening of the major wind patterns in this area leads to alterations in the surface currents in the southern gyre, suppression of the Peruvian upwelling and changes in the region's biological productivity, sea level and rainfall patterns.

These changes in the southern Pacific have considerable effects on the patterns of rainfall and weather in Australia and Southeast Asia, where monsoon rainfall is reduced during El Niño years. The phenomenon is called El Niño (Spanish, 'the Christ child') since it occurs in Latin America around Christmas and forms part of the ENSO (El Niño, Southern Oscillation) pattern of climate in the southern hemisphere. The impacts of such events can be traced in patterns of coastal dune formation in eastern South America and may be implicated in changed rainfall patterns as far away as Africa.

The Southern Gyre

Under normal conditions, the South Pacific Gyre dominates water circulation in the southern Pacific Ocean. Surface water flows eastward under the influence of the trade winds as the South Equatorial Current, passes down the eastern coast of Australia (as the East Australia Current) and turns eastward at the latitude of Sydney under the influence of the westerly winds. The Antarctic Circumpolar Current then moves eastward toward Latin America where a major branch, the Humboldt Current, travels northward along the coast of Chile toward Peru.

The high Andes block the passage of the westerlies, deflecting them northward, which not only adds to the speed of the Humboldt Current, but also causes the surface waters to move away from the coast. As they do so, they form part of the eastward-moving South Equatorial Current, and subsequently, cold, nutrient-rich water is drawn to the ocean surface.

▼ ***The left-hand globe*** *shows the effect on the ocean climate following the 1997–98 El Niño and the 1998–99 La Nina. Unusually cool water (areas of lower sea level shown in blue and purple) extends from the Gulf of Alaska along the coast of North America, sweeping southwest from Baja California, where it merges with the remnants of La Nina. Areas where the temperature of the Pacific is normal appear in green. The right-hand globe shows the equatorial Pacific in June 2000, warming up and returning to normal after the two-year La Nina episode. Only a few patches of cooler, lower sea levels remain across the tropics.*

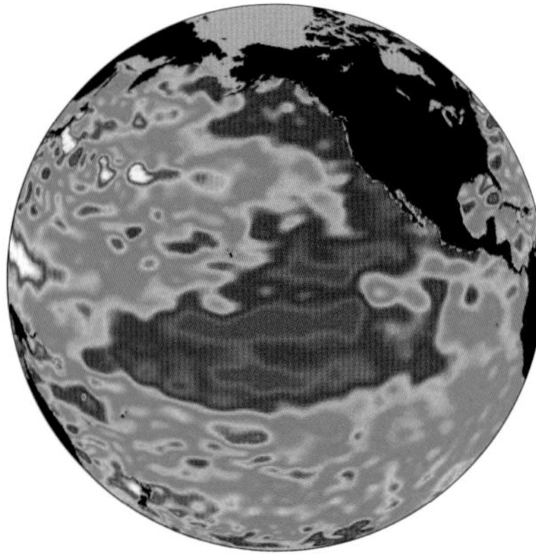

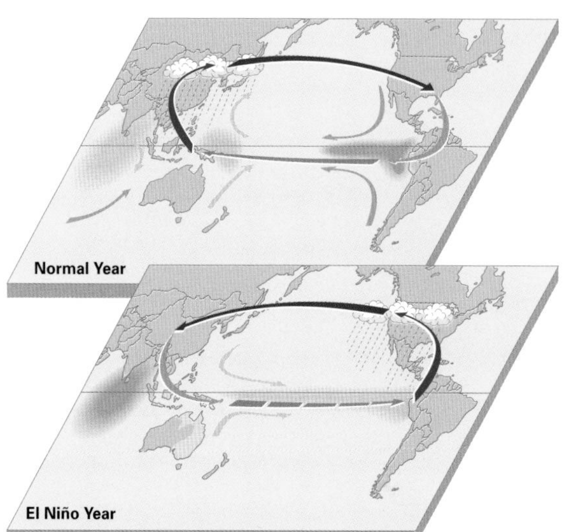

► ***In a normal year*** *in the Pacific Ocean (top), southeasterly trade winds drive sun-warmed surface waters westward off the coast of South America into a pool off northern Australia. High cumulus clouds form above these warm waters, bringing rain in the summer wet season. Cooler, nutrient-rich waters rise to the suface off South America, supporting shoals of anchovetta on which the fishing industry depends. Every three to seven years, a change occurs in the ocean-atmosphere interaction. The climatic pattern is reversed (bottom map) – an event known as El Niño. The trade winds ease and the warm surface waters in the western Pacific flow back to warm the waters off South America by 4–5°F (2–3°C), or even as much as 12°F (7°C). The warm waters off South America suppress upwelling of the cold nutrient-rich waters, spelling disaster to the fishing industry. During an intense El Niño, the southeast trade winds reverse direction and become equatorial westerlies, resulting in climatic extremes.*

The Peruvian upwelling

Off the coast of Peru, this upwelled, cold, nutrient-rich water stimulates the production of phytoplankton and a characteristic community of comparatively large, multicellular phytoplankton develops. The size of these primary producers, when compared with the smaller, single-celled diatoms characteristic of nutrient-poor, open ocean water, means that they can be eaten directly by small fishes, in particular the valuable anchovetta.

The anchovetta themselves form the food of larger predatory fishes and numerous seabirds, whose breeding colonies on the Latin American coast have resulted in the deposition of guano, mined during the last century as phosphate fertilizer. In addition, the anchovetta are extensively fished from small purse seiners, and the catch processed to produce fishmeal for domestic animal and poultry feed. The high fisheries production of the Peruvian upwelling reflects both the high rates of primary production in the area and the efficiency of the food chain, which has a reduced number of linkages between the primary producers and the top predator, in this case man.

Physical changes

During an El Niño event, upwelling ceases and the biological productivity of the area collapses, since the important nutrients are no longer brought to the surface. Extensive research on this phenomenon during the last decade has shown that the surface currents of the Southern Gyre are weaker during El Niño years, and the eastward-moving South Equatorial Current becomes dominated by the westward-flowing Equatorial Countercurrent.

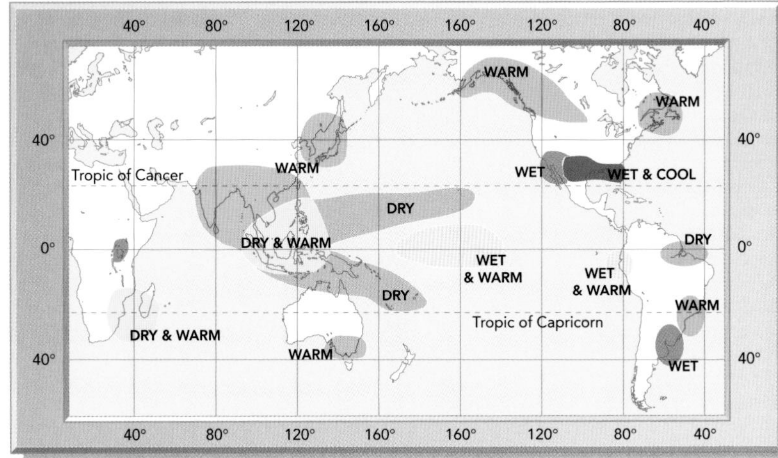

▲ ***Typical thermal*** *and rainfall anomalies (December to February) during a strong El Niño.*

During its 9000-mile (15,500-kilometer) journey across the Pacific Ocean, the surface waters of the South Equatorial Current become progressively warmer, leading to the formation of a pool of warm water in the western Pacific. In addition, the movement of the water results in higher sea levels in the western Pacific region and lower sea levels along the Latin American coast. When the flow weakens or is reversed, the water levels change on each side of the Pacific Basin. During El Niño years, mean sea level in the western Pacific may be as much as 5.5 inches (14 centimeters) below normal level, while along the Latin American coast it may be as much as 20 inches (50 centimeters) higher than normal. The surface waters in the zone of upwelling are now dominated by the warm waters from the western Pacific, upwelling ceases and the biological productivity of the region collapses.

Economic effects

The pelagic fishery associated with this productive zone of upwelling is composed largely of three species – the anchovetta, sardine and jurel – and the fishery contributes more than 16 percent of the total fish catch in the Pacific. The potential sustainable yield of the anchovetta fishery is estimated at 12.6 million tons or around 9 percent of the total world catch. In 1970, the Peruvian and Chilean catch of anchovetta amounted to around 13 million tons, but this fell to 1.2 million tons during the 1972–73 El Niño.

In contrast, some mollusks, such as the Peruvian scallop, become more abundant during El Niño events, since spawning and survival are better under warmer conditions. The 1983 catch of this species reached 10,000 tons compared with normal yearly landings of around 1000 tons.

During the El Niño events in 1972–73 and again in 1982–83, the catch of anchovetta was reduced to one-sixth and one-six-hundredth of the normal catch, respectively. As the fish stocks decline, the birds, which depend on the anchovetta for food, also die in large numbers. Following the 1972–73 El Niño, the populations of cormorants, boobies and brown pelicans declined to around 6 million birds, from an estimated number of 30 million in 1950. The 1982–83 event reduced their numbers to a level of only 300,000. The 1997–98 El Niño was the strongest on record. It killed an estimated 2100 people, and caused at least US$33 billion dollars in property damage.

The economic consequences of El Niño are severe. Raised water levels along the Latin American coast increase the frequency of malaria, while inland flooding results in disruption of agriculture and the destruction of coastal infrastructure. Increased rainfall in the coastal regions contributes further to these problems. Economic impacts also occur elsewhere, with agriculture being adversely affected in the semiarid regions of Australia and subsistence food production being reduced in Southeast Asia.

Related changes

The El Niño phenomenon appears to be related to changes in atmospheric circulation and in particular to a weakening of the westerly winds which drive the southern arm of the South Pacific Gyre. The consequences of such a major change in the air-sea interaction are not confined to the Pacific Basin. These interactions appear to be linked to other changes in atmospheric and oceanic circulation in the Southern Hemisphere. The results of these changes are reduced rainfall in Australia, weakening of the Indian monsoons, changes in the Kuroshio Current off Japan, shifts in wind patterns in the South Atlantic, with consequent changes in the orientation of coastal dunes in Argentina, and droughts in Southeast Asia.

► ***Pelicans and gannets*** *on a fishing boat net. Seabirds like these are directly affected by the changes in marine productivity in El Niño years.*

Ocean Exploration

EARLY SEAFARERS

For thousands of years people have had a close relationship with the oceans and their resources. Unfortunately, much of the early archaeological evidence of this relationship has been lost following the rise in sea level at the end of the Pleistocene period about 10,000 years ago. Early coastal habitation sites some 40,000 years old are, however, known from accumulated food refuse in southern Africa and Melanesia.

Coastal regions offered settlers the combination of fertile agricultural land and abundant protein in the form of shellfish and finfish. Extensive shell middens (areas of accumulated refuse) along old coastlines worldwide indicate that gleaning, or the collecting of food, from the intertidal area has long been important to human societies.

Early craft

The first boats were almost certainly those built for use on rivers, lakes and coastal swamps. The bark canoes of the North American Indians and the skin-covered kayaks of the Inuit, for example, were all in wide use at the time of first European contact, while on the Sepik River of New Guinea, canoes carved from logs, often with ornately decorated prows, are still a common sight. Skin-covered boats, known as coracles or curraghs, were also in use in France and Ireland until relatively recently.

Such boats, although light and manoeuvrable, lack sufficient stability for use in rough seas. Early Pacific islanders discovered that fitting an outrigger to a single-hulled canoe greatly increased its stability. This innovation, together with the addition of sails rather than relying on paddles or oars, enabled the islanders to venture further offshore in search of migratory fish, such as tuna.

▼ ***Ra II**, with a crew of seven captained by Thor Heyerdahl, puts to sea off Safi, Morocco, in an attempt to prove that Atlantic Ocean currents could have carried reed ships from Africa to the Americas in ancient times. Ra II successfully completed the voyage from Morocco to Barbados in just 57 days.*

The first seafarers

Although we have no information on the earliest ocean-going vessels, we know people successfully crossed from southeast Asia into Australia and New Guinea at least 40,000 years ago. Archaeological remains from Australia and New Guinea indicate that communities had been established by this time at widely different locations. To have achieved this, large numbers of people must have crossed from southeast Asia, a voyage that would have involved, even during the periods of lowest sea level, an open stretch of ocean at least 25 miles (40 kilometers) wide.

Seagoing craft were developed independently in many different parts of the world. One of the oldest surviving ships is the Egyptian Cheops ship, a planked craft dating back to 3000 BC. The colonization of the islands of the central Pacific, which occurred around 4000 years ago, depended on the building of ocean-going craft.

In Europe, maritime trade was well established as long as 5000 years ago, when seafarers from the Aegean traded in obsidian around the Mediterranean. The founding of colonies on Greenland and Iceland, and the discovery of America by the Vikings, attest to the versatility and seaworthiness of their advanced longboats.

In the Indian Ocean, the development of trade links stretching from the coast of East Africa to Arabia and the western Indian coastline, must have developed quite early, since the Maldives and Sri Lanka are known to have sent emissaries to Rome more than 2000 years ago. By the time Europeans penetrated southeast Asia, the Arabs had established an extensive trading network based on colonies and outmails which stretched from the east coast of Africa to the tip of New Guinea.

▲ ***Arab dhows*** *were in use more than 1300 years ago. They made long transoceanic voyages to the East African coast, India, Southeast Asia and China, carrying Arab and Persian traders to Guangzhou (Canton), southern China.*

Ocean pathfinders

Although we are aware of early navigational instruments, such as the stick charts of the Micronesians which were used to great effect some 1000 years ago, the earliest ocean voyagers most likely remained close to land, sailing along a coastline for orientation, or for short distances out at sea to use favorable winds. European ocean voyaging is relatively recent, and the European dominance of the world's oceans is relatively short in terms of human history, no more than 500 years.

▲ ***Sixteenth-century map of Africa*** *by Giorgio Sideri. Althougl pilots could calculate their latitude at sea, longitude could not be measured accurately, hence good maps, sailing directions and skilled pilots were essential for maritime voyages.*

This dominance of ocean trade and transportation stems from the extensive investment in voyages of exploration, which characterized European history during the 15th and 16th centuries. Sent out by Prince Henry in the 1420s to discover a maritime route to the source of African gold, Portuguese seamen discovered the Azores and Madeira, and in 1434 rounded Cape Bojador at 26°N – the previous, southern limit to Atlantic seafaring. To achieve this, it would have been necessary to sail south using the trade winds, then due west to encounter winds from the south and then use westerly winds to bring them back to Portugal. The Arabs used similar wind patterns in the western Indian Ocean in their trading voyages.

Early navigation

The early voyages of exploration were based on the use of compass and sandglass. The compass provided mariners with direction, while the sandglass was the standard marine timepiece and was used with log and line to measure distances. These crude but effective instruments, together with written directions, had allowed reliable year-round navigation of the Mediterranean region for some considerable time. The ocean-current systems of the Atlantic, however, made calculations using only these navigational aids inaccurate.

The development of nautical astronomy brought about a major breakthrough in ocean navigation. By measuring the altitude of the North Star and later the Sun at noon, and multiplying by the number of miles in a degree, a skilled pilot could provide a reliable estimate of the distance sailed in a northerly or southerly direction. By 1480, astronomical rules and tables had been developed allowing pilots to calculate their latitude. In 1488, using these sorts of instruments and tables, the explorer Bartholomeu Diaz rounded the Cape of Good Hope and returned to Lisbon having discovered a sea route to the East. Ten years later, Vasco da Gama, using the anticlockwise wind circulation of the South Atlantic and Indian Oceans, reached India in his two specially constructed three-masted, square-rigged ships. Such ships were developed in Europe around 1450 and, armed with cannon, gave Europe mastery of the oceans for nearly four centuries.

Age of discovery

At the beginning of the 16th century, the further development of navigational aids enabled mariners to determine their position at sea with increasing accuracy. The sea astrolabe provided precise measurements of the altitudes of the stars and planets, and when used in conjunction with the plane charts, which provided a latitude scale, pilots were able to sail north or south to the latitude of a known landfall before sailing due east or west to make land. Columbus used this navigational technique for his return after crossing the Atlantic in 1492 to discover the 'New World'.

During this period of exploration, the Spaniards in Lisbon and Seville continued to chart discoveries and develop navigational techniques, and by 1516 the first world map based on these discoveries was published by Waldseemüller. The vast extent of the Pacific was only discovered during the circumnavigation of 1519–22, led initially by Magellan and completed by Del Cano.

Meridional tables published in 1599 by the mathematician Edward Wright enabled hydrographers to construct charts mathematically on what became known as the Mercator projection, after the cartographer Gerhardus Mercator. The first sea chart to use these tables was also published in 1599, and theoretically allowed a ship's position to be plotted in terms of latitude and longitude and could provide an accurate chart of the distance and direction to landfall. The problem

◀ ***Columbus' ships*** *(the* Nina, *the* Pinta, *and the* Santa Maria) *at sail on the high seas. The* Santa Maria *was the flagship of the fleet and led the 15th-century expedition that would eventually discover America.*

▶ **Endeavour** *from* Ships that Made History *(1936) by Geoffrey Robinson. Captain Cook sailed the Endeavour on his voyage of 1768–71. He carried observers to Tahiti to watch a transit of the planet Venus in front of the Sun. The voyage was also to explore the southern oceans in search of a great southern continent. Cook instead discovered New Zealand, charting its coast for six months, before finding southeastern Australia in April 1770. Cook charted Australia's east coast and navigated unscathed through the Great Barrier Reef. He was famous for insisting on a fruit-rich diet to prevent scurvy. In 1778, on his third voyage, Cook was killed by Hawaiian islanders.*

of accurately determining longitude at sea was not solved, however, until much later and required the development of precise navigational instruments.

Navigational instruments

Accurate timekeeping and variations in the Earth's magnetic field caused navigators problems, and several attempts were made to improve on the timekeeping of the sandglass. John Harrison's marine timekeepers were essential developments for determining longitude. Captain James Cook used a copy of his fourth chronometer during his voyages of exploration. Although magnetic variation had been known to compass makers since the 1450s, it was believed to be constant wherever it was observed. In 1635, the annual change in the Earth's magnetic field was discovered and in the late 1690s the British Admiralty commissioned an astronomer, Edmond Halley, to measure the variations in the North and South Atlantic. He published the first isogonic (lines of equal variation) chart in 1701, later extending it to the Indian Ocean.

From the late 15th century onward, Portuguese navigators determined their latitude by observing the Sun's or a star's altitude at noon with an astrolabe, while the cross-staff, developed around 1514, allowed the Sun's altitude to be read directly. Although latitude could be determined with relative accuracy during the 16th century, the determination of longitude was not possible until the 18th century.

Greenwich Observatory

Although Galileo, who invented the pendulum-controlled clock and astronomical telescope, discovered that the satellites of Jupiter could be used for determining longitude with reasonable accuracy, the method proved to be impractical for use aboard a ship. In 1675 the Royal Observatory was founded at Greenwich, London, with the charge of solving this problem.

In 1760, John Harrison's fourth chronometer and the lunar-distance method of calculating longitude provided the solution to the problem. The publication of the annual nautical almanac from 1767, which gave the lunar distances from the Sun and certain stars for every three hours at Greenwich, combined with John Hadley's invention of the reflecting quadrant in 1731, subsequently improved into the sextant in 1757, enabled lunar observations to be made on board ship. The sextant works by reflecting an image of the Sun or a star, via a mirror, onto a glass plate through which the horizon is visible. By adjusting a calibrated, sliding arm, the angle between the Sun or star can be altered until the image rests on the horizon, so determining the longitude.

During his three voyages between 1768 and 1780, Captain Cook explored the Pacific Ocean using the lunar-distance method for determining longitude, and his accurate charts provided the first basis for the economic exploitation of this vast expanse of the world's surface.

▼ ***The Royal Observatory*** *at Greenwich, London, from Ackermann's Repository of Arts (c.1826). In 1675 Sir Christopher Wren, commissioned by King Charles II, designed the facade of the building. The Royal Observatory is famous as the site of the meridian at 0° longitude, Greenwich Mean Time (GMT), from which all world time zones are measured.*

OCEANOGRAPHY

Although the Greek philosopher Aristotle studied the marine life of the Aegean and discussed various theories concerning the salinity of the sea, navigational problems and the lack of suitable equipmentthe early exploration and study of oceans was hampered by .

Early development

In the 17th century, Sir Isaac Newton used tidal information, compiled from seafarers' observations collected in the pamphlet *Directions for Seamen* relating to the depth of the sea, to illustrate his theory of gravitation. While around the same time, eminent scientists such as Robert Hooke produced designs for depth sounders, water samplers and deep-sea thermometers.

During the early 19th century, Alexander Marcet studied the salinity of the world ocean, James Renell produced charts of Atlantic Ocean currents, and Emil von Lenz studied variations in temperature and salinity in the deep ocean, confirming that density differences in the ocean are responsible for ocean currents.

Since 1855, when Matthew Fontaine Maury published the first chart of the North Atlantic Ocean floor, the science of bathymetry (the study of the physical form of the ocean floor), has progressed significantly, and accurate bathymetric charts for much of the world's ocean floor are now widely available.

Oceanography

Until the middle of the 19th century, most scientists believed that life could not survive below around 400 fathoms – 2300 feet (700 meters) – but in 1869, Wyville Thompson dredged creatures from 2500 fathoms – 15,000 feet (4600 meters) – during the voyage of HMS *Porcupine*. The voyage of HMS *Challenger* (1872–6) was considered the first truly oceanographic cruise, and many nations followed this early example. Studies of deep-water circulation by German ships led to the first accurate model of circulation in the Atlantic.

By 1900, the study of the oceans had become recognized as an important area of scientific endeavor, although many early institutions were little more than seaside stations which specialized in biology; some, such as the institutes established in Paris and Monaco, were more diverse in their research interests. One of the first areas of marine research was the fishing industry, resulting from concern as early as the 1870s over the decline of fish stocks, particularly those of the North Sea. By 1925, Britain had established the 'Discovery Investigations'

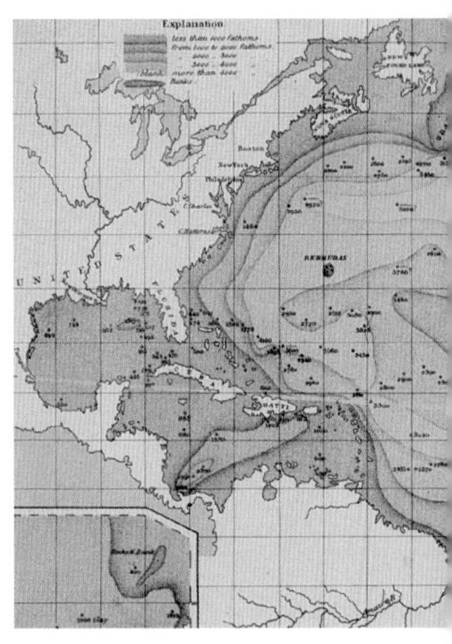

▼ ***The first map*** *to be made of an ocean basin charted the bed of the North Atlantic. By Matthew Fontaine Maury, it appeared in his book* The Physical Geography of the Sea. *Maury used soundings made with a lead and line to prepare his map.*

The *Challenger* expedition

One of the most important scientific voyages ever made, the *Challenger* expedition of 1872–76 laid the foundations for the modern science of oceanography. Two biologists, W.B. Carpenter and Wyville Thomson persuaded the British government to equip the expedition to study deep-sea circulation and the distribution of life in the seas. The voyage was the first to discover manganese nodules on the ocean floor, which were found at all sites in the ocean basins. The expedition also sampled fauna and fishes down to depths of 4500 fathoms – 27,000 ft (8000 m) – thus demonstrating the existence of life in the abyssal depths.

▲ **Challenger's *expedition*** *report was illustrated with diagrams of organisms, such as these drawings of sponges.*

◀ ***The* Challenger *expedition*** *carried a team of scientists on its four-year circumnavigation.*

in response to concern about the decline in whale stocks in the Southern Ocean.

During World War II, pioneering work on wave forecasting and underwater acoustics was undertaken as war was taken beneath the sea surface by submarines. Since that time, oceanography has become a sophisticated and highly technical science, dependent on a number of devices, which permit scientists to examine and measure areas directly inaccessible to human observers. Satellite observations of the sea surface, for example, enable simultaneous estimation of primary production over vast areas and provide extensive data on surface water temperatures, wave and current patterns.

Commercial vessels routinely traversing the ocean basins participate in a coordinated 'ships of opportunity' program, releasing instruments to measure temperature and salinity profiles at different points along their navigational route. Observations of the marine environment at depth are now possible through the use of remotely controlled, deep-diving equipment. The spectacular rediscovery of the *Titanic* and observations of the diverse marine communities of submarine thermal vents, bear witness to our rapidly expanding knowledge of the ocean realm.

▲ ***Edmund Halley's*** *diving bell of 1690 was the first bell in which divers were not restricted to the amount of air contained in the vessel.*

People have long been fascinated with the possibilities of underwater vehicles. Attempts to produce such craft date back to the 4th century BC, when Alexander the Great is believed to have descended beneath the Mediterranean Sea in a vessel resembling an early diving bell.

Early submersibles

The first use of a submersible was in 1776, when a small submersible, the *Turtle* built by David Bushnell, was used against the British in the American Revolution. In 1863, the submersible *David* was used successfully in the American Civil War, actually sinking an enemy ship. Thirty years later, Simon Lake constructed the *Argonaut First* – a true submersible in which atmospheric pressure could be maintained, and the vehicle moved around the seabed on hand-powered wheels. Following these early vehicles, many navies developed submarines. Conventional submarines use diesel power on the surface and battery power while submerged. In contrast, nuclear submarines, powered by means of a nuclear reactor, are capable of remaining submerged for months at a time.

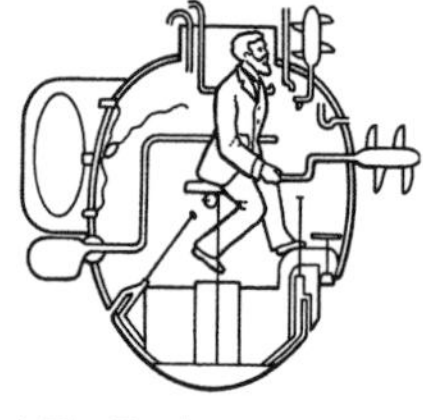

▲ **The Turtle** *was a one-man submersible built in 1776. It was powered by propellers and was used in attempt to sink a British warship during the American Revolution.*

Exploring the depths

Military submarines are generally unable to descend to depths of more than a few hundred meters, and the breakthrough in submersible exploration of the deep ocean regions occurred in 1934, when William Beebe and Otis Barton descended to a depth of around 3017 feet (920 meters) in Barton's revolutionary bathysphere. The bathysphere was a heavy steel sphere that was lowered into the ocean on a cable payed out from a surface support ship.

A later version, the bathyscaphe invented by Professor Auguste Piccard, was free of surface cables and the bathyscaphe *Trieste* was used by Piccard for a dive of 10,392 feet (3170 meters) off the coast of Italy in 1954. In 1960 Piccard's son Jacques led a dive in the Marianas Trench in the southwestern Pacific, touching the bottom at a depth of approximately 35,800 feet (10,900 meters), a record yet to be exceeded.

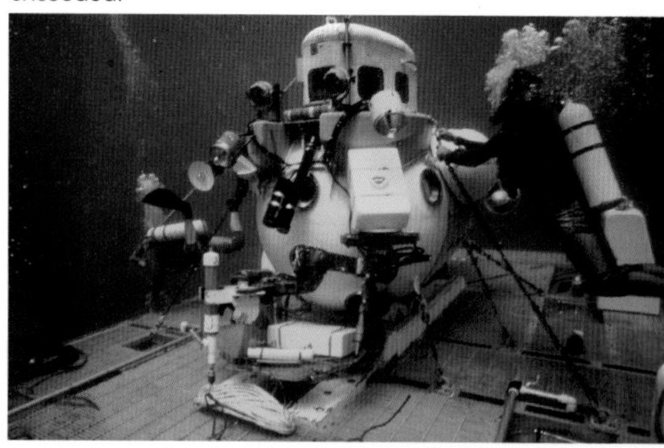

► ***Divers secure*** *the* Makali'i *sub to the LRT submersible barge before surfacing. The* Makali'i *is operated and maintained by the Hawaii Undersea Research Laboratory.*

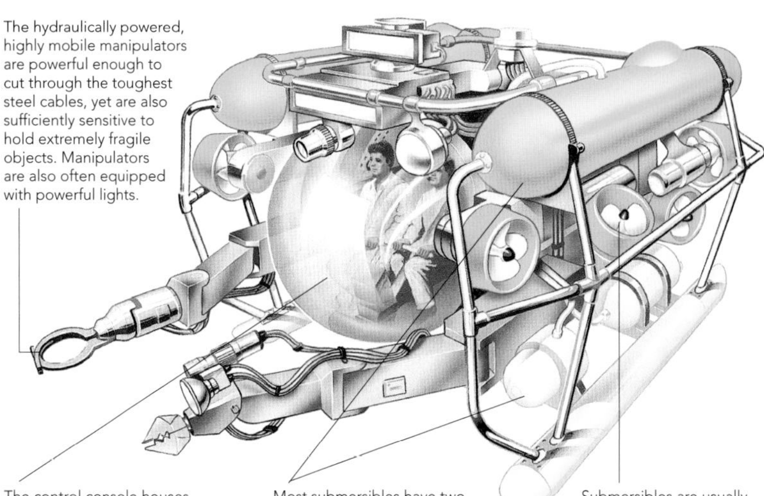

The hydraulically powered, highly mobile manipulators are powerful enough to cut through the toughest steel cables, yet are also sufficiently sensitive to hold extremely fragile objects. Manipulators are also often equipped with powerful lights.
The control console houses sophisticated computer-aided navigational and recording equipment. With the sort of technology available, some submersibles are capable of working either manned or unmanned.
Most submersibles have two buoyancy systems. The larger cylinders are used to raise and lower the craft to and from the main operating area, while the lower, smaller tanks are used to make finer depth adjustments.
Submersibles are usually fitted with a number of directional propellers known as thrusters. These give the craft great maneuverability when working under water.

▲ ***Modern submersibles****, such as the* Johnson Sea Link *shown here, are versatile vehicles used for a variety of scientific and industrial purposes. Most can be fit with a wide range of equipment, cameras and clawlike manipulators.*

Scientific submersibles

One of the earliest and most successful of the scientific submersibles was the American *Alvin*. First commissioned in 1964, *Alvin* was capable of diving to almost 2 miles (3 kilometers) and was later upgraded so that it could achieve depths of 2.5 miles (4 kilometers). *Alvin* submersibles have made thousands of highly successful and productive dives, but none more spectacular than the early *in situ* observations of hydrothermal vents in the 1970s. The more recently developed scientific submersibles such as the American *Sea Cliff*, the French *Nautile*, the Russian *Mir I* and *Mir II*, and the Japanese *Shinkai 6500* are all able to reach a depth of 3.7 miles (6 kilometers), and thus penetrate all ocean depths except for the deepest ocean trenches.

Remotely operated vehicles (ROVs)

ROV craft are mostly small, manoeuverable units capable of transmitting television pictures or other information and are usually controlled via an umbilical cable from the surface or from a second submersible, either manned or unmanned. Perhaps the best known of these systems is the combination of the towed, remotely operated vehicle *Argo* and the tethered small vehicle *Jason Jr.*, which were used by Dr. Robert Ballard of the Woods Hole Oceanographic Institute for the dramatic investigation of the wreck of the *Titanic* in 1985. By late 1992, there were 194 heavy-duty ROVs in use in the offshore oil industry worldwide.

PROBING THE DEEP

In 1872, HMS *Challenger* undertook a four-year voyage that marked the birth of modern oceanography. The ship carried numerous new oceanographic instruments, many of which were unproven but which were basically similar in design to today's water samplers, corers, dredgers and current meters. Oceanographic sampling instruments are designed either to bring back samples of water, sediment or animals from depths, for subsequent study and analysis, or to record the physical conditions beneath the surface.

Most water-sampling devices consist of a cylinder, open at both ends, which is lowered to the required depth. A 'messenger', a weight, is dropped down a guideline to release the two ends which snap shut to close the cylinder and prevent the water from being contaminated as it is drawn to the surface. Similar mechanisms are used to open plankton nets in order to sample the planktonic organisms at different depths. More sophisticated nets may incorporate a flow meter at the mouth to measure the volume of water which has passed through the net during the sampling period. By counting the number of individuals and relating this to the volume of water, the density of organisms, and hence the productivity of particular water bodies, can be estimated.

Continuous plankton recorders are towed behind moving vessels and trap the plankton on a continuously moving roll of gauze, which passes into a tank of preservative. Relating the samples to the ship's log enables the samples to be located along the path of the ship's voyage. Ships towing these samplers through the North Atlantic over many decades, have provided information about changes in the productivity of the ocean, the onset of spring blooms of plankton, and the role of water temperature in the annual cycle of productivity.

▶ ***Box corer*** *operating on the Deep Ocean Mining Environmental Studies (DOMES) project in the Pacific.*

▼ ***The Hardy Continuous Plankton Recorder*** *is towed at a depth of 30 ft (9 m). The flow of water past the propeller drives a series of rollers which wind a continuous strip of gauze. Plankton enter the front and are trapped on the gauze before passing into a tank of preservative. The gauze is unwound and in conjunction with the ship's log provides a continuous record of the plankton through which the ship passed. Records from such instruments provide data on seasonal cycles and longer term changes in plankton in the North Atlantic.*

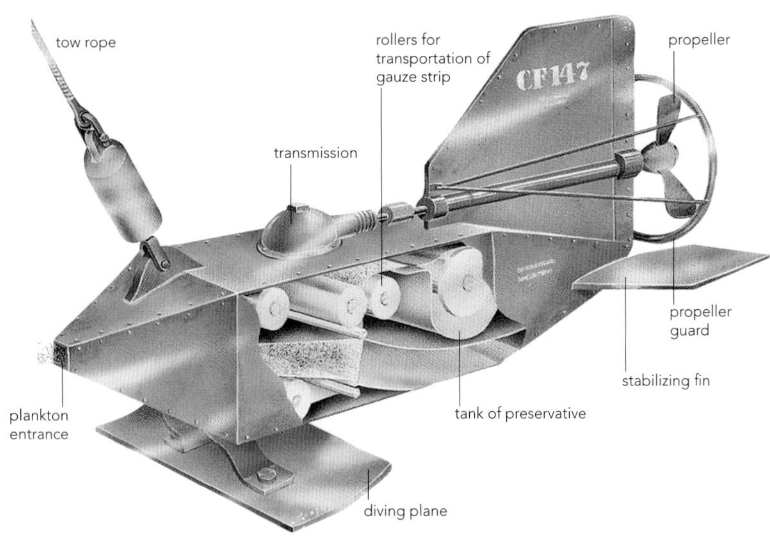

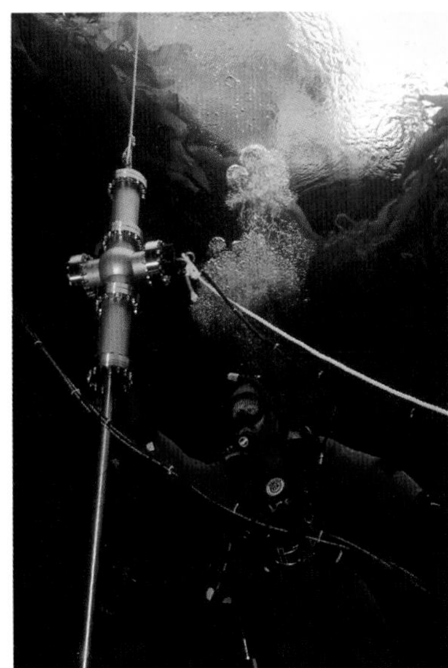

▲ ***A new aluminum*** *deep sea probe, a prototype designed to withstand crushing pressures and extreme temperatures, is lowered to depths of 30 ft (9 m) in Monterey Bay Aquarium's giant kelp forest (California, USA) as part of NASA's hunt for clues to the origins of life.*

Samplers and corers

Seafloor samplers include a variety of trawls and dredges which are towed along the bottom to collect surface samples of sediment and rocks, and benthic animals. This technique is destructive and frequently results in damaged specimens. More sophisticated grabs are dropped, with jaws open, from the ship; the jaws close as the grab hits the bottom. Some can be operated independently and have weights causing them to sink, which drop off as the sampler jaws close. Flotation chambers then cause the sampler to rise to the surface.

Seabed samples are taken by means of corers, the simplest of which are dropped over the side of a ship. The core box penetrates the sediment, and a device closes the end of the corer ensuring that the sample remains intact as the device is hauled to the surface. The development of drilling technology associated with oil exploration has greatly improved sampling techniques, and the *Glomar Challenger*, a research drilling ship, has retrieved cores from 4232 feet (1290 meters).

On-site recorders of physical parameters range from simple reversing thermometers used in conjunction with water sampling devices to record water temperatures, to expendable bathythermographs which measure a range of variables, including temperature, pressure and salinity. Drifting buoys are employed to study current systems. Some float on the surface and their signals, relayed to satellites, provide information on the strength and direction of currents. Others operate using floats which maintain them at a precise depth, providing information on underwater currents.

The earliest maritime law was the Rhodian Sea Law or Rhodian Code, which dates to the 3rd or 2nd century BC. Devised to apply to the Mediterranean, its principles were adopted by both the Greeks and Romans and observed for 1000 years. In the early 17th century, Hugo Grotius published the treatise *Mare Liberum*, which formed the basis for the subsequent adoption by all maritime nations of the concept of the 'Freedom of the Seas' and recognition of the territorial sea, covering waters within 5 to 3 to 6 miles (10 kilometers) offshore. Territorial waters were considered to be under the jurisdiction of the coastal state, while international waters were open to individuals of all states for navigation and fishing. Thus seafarers traveled extensive distances in search of fish, whales and seals without thought for ownership of the resources concerned.

► ***In the upper diagram**, changes in forestry affect the mangrove habitats of juvenile shrimps thus affecting the trawl fishery. In the lower diagram, several neighboring countries share a transboundary stock of tuna which is also fished by distant fishing fleets on the high seas. Management of the tuna stock is impossible unless all countries agree. In the case of the diagram for shrimp, fishermen have no control over land-use changes affecting their resource.*

The United Nations

In the 19th century, overexploitation and the decline in many fish stocks led to attempts to regulate fish catches based on voluntary quota systems. These often failed because states that did not agree with the quotas, simply withdrew from the fisheries commissions or failed to sign the agreements.

In 1930, the League of Nations tried to secure international agreement to the declaration of the 3-mile (5-kilometer) territorial limit. Discussions broke down and in 1945 President Truman issued a unilateral proclamation of the United States, declaring exclusive right to exploit its continental shelf. This declaration was soon followed by the declaration of an exclusive 200-mile fishing zone by the Pacific States of Latin America. In 1958, the United Nations convened the first UN Conference on the Law of the Sea. The Conference

► ***The boundaries** between legal and administrative units rarely coincide with the scale at which processes occur in the ocean environment.*

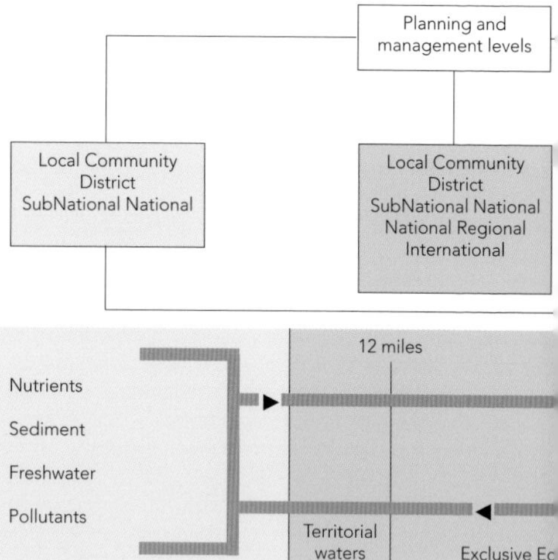

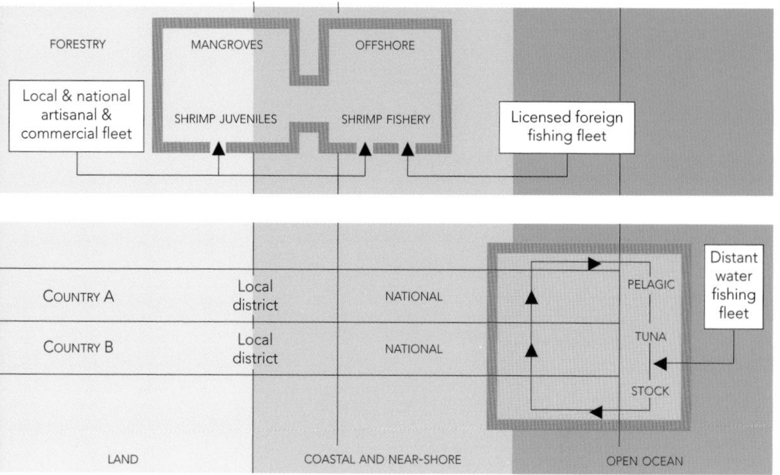

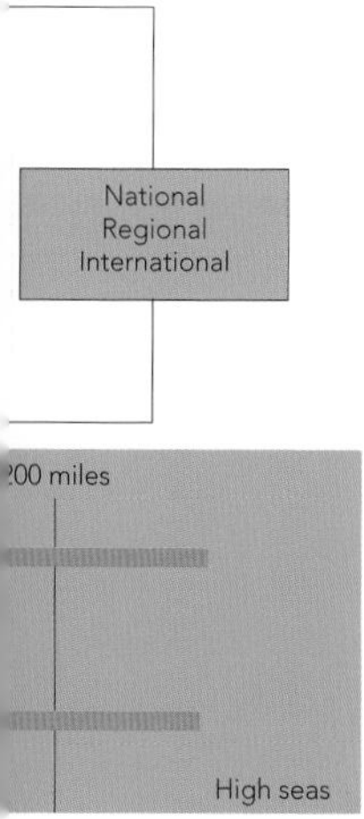

established the principle of a 12-mile contiguous zone within which the coastal state was permitted to enforce customs, sanitary and fiscal regulations.

During the 1970s, it was proposed that no state should extend its unilateral control beyond the limits of the territorial seas, and the General Assembly of the United Nations called for a moratorium on exploitation of the minerals of the seabed beyond the continental shelf. A declaration of principles was unanimously adopted proclaiming the seabed "the common heritage of mankind". In 1973, the United Nations convened the third Conference on the Law of the Sea with the aim of developing a single treaty encompassing the intentions of existing international conventions. Following numerous sessions held over a period of years, a single text was finally adopted in 1982 and the Convention became binding on all parties on November 16, 1994. A majority of the 130 coastal states have adopted the 12-mile or lesser limit for their territorial seas, and 91 states have declared their 200-mile Exclusive Economic Zones (EEZs), within which states have the right to exploit both living and nonliving resources but over which they have no territorial rights.

Other treaties include: 11 Regional Seas Conventions, negotiated between states with a shared area of ocean space, such as the Mediterranean Sea. These include the London Dumping Convention that regulates ocean dumping; the MARPOL Convention that regulates ship-based discharges; and the conventions regulating fisheries based on shared, transboundary stocks. The need for this increasingly complex system of regulations at the national, regional and international level reflects the increasing pressure which is resulting from the unregulated use and abuse of marine resources.

National
Regional
International

200 miles

High seas

Ocean Leisure

SURFING

The first recorded mention of 'surfing' is in the account of the 1779 landing of HMS *Bounty* at Tahiti, now French Polynesia, where men, women and children were seen riding the waves on short wooden planks. In Hawaii, the sport was a part of everyday life, with strong class overtones, since the longest boards and best surfing breaks were reserved for the nobility. The long boards used by the nobles were called *olos* and reached 16 feet (5 meters) in length and weighed up to 150 pounds (67 kilograms). The commoners used shorter boards, *akaias*, that were up to 8 feet (2.5 meters) long, while small children used short planks called *paipos*, similar to modern bodyboards.

Following the influx of Europeans to Hawaii in the mid 1800s, surfing declined – in part due to the influence of missionaries and in part due to the breakdown of traditional culture under the onslaught of disease, Christianity and

◀ ***Commander William M. Scaife*** *of the United States Coast and Geodetic Survey with an early surfboard, Waikiki beach, Hawaii, 1926. Developments in board design made the surfboards lighter and easier to use.*

▲ ***A surfer** catches a big wave off the coast of Hawaii. Surfing is now a major international sport with competitions held all around the world. Professional surfers can earn thousands of US dollars for high-ranking places.*

commercial development. By the start of the 20th century, only a few Hawaiians were still surfing but gradually interest in the sport grew as nonHawaiians took to the surf. Revival of the sport was greatly influenced by the Hawaiian Duke Kahanamoku, who competed as a swimmer in four Olympic Games from 1912 to 1928, and who together with his brothers formed surfing clubs in Hawaii to encourage interest in the sport. In the early 1900s George Freeth, an Irish-Hawaiian, gave surfing exhibitions on the US mainland, stimulating interest in surfing on the beaches of California, and the revival of the sport began in earnest.

Surfboard design

One of the first changes that came to surfboard design was stimulated by the desire for lighter more manoeuvrable boards, and by the late 1920s hollow boards without fins were in use. Later balsa was used, edged with harder redwood along the nose, edge and tail of the board. With the addition of a fin, the lightweight balsa boards were shorter and could be turned more easily in heavy surf. The 1940s saw the introduction of new materials, like fiberglass cloth, while plastic resins replaced the varnish used on balsa boards, giving the board greater durability and strength.

Balsa boards required considerable skill in shaping and working the wood and quickly became waterlogged if the protective varnish or fiberglass coating became cracked. In 1946 the foam-filled board was designed, which became the forerunner of all modern boards. By 1958 the first fully foam board was created which revolutionized surfing. Boards were now shorter, lighter and more manoeuvrable, and could be more easily carried and paddled, enabling more women to take up the sport.

Surfboard types

There are three basic types of surfboard – the traditional long board with a rounded nose, the board with a straight midsection, and the board with a wide tail. Longboards can be 8 to 12 feet (2.4 to 3.6 meters) in length and generally have a single fin on the underside near the tail. The narrow pointed shortboards are generally thin and streamlined with an abrupt rise in the nose and can be 5 feet, 6 inches (14 centimeters) or longer with three fins. Hybrids combine elements of both the long and short boards, and are commonly called funboards.

In the 1970s, flexible foam bodyboards became common, being easier to ride, costing much less than surfboards and being much more portable. Bodyboards are ridden either lying flat or with one knee raised and they represent an addition to the traditional bodysurfing which involves riding a wave to shore with no board and nothing separating the swimmer from the wave itself.

WINDSURFING

Windsurfing began in the late 1950s, and the first windsurfing craft was patented in the late 1960s. During the 1970s, hundreds of thousands of windsurfing boards were sold and the sport became a craze in Europe, so much so, that by 1984, only some 15 years after the first patent, the sport was included in the Summer Olympics. Windsurfing evolved quickly, with racers in Europe building big boards for light wind racing and the wave sailors of Hawaii building small

▼ ***A colorful group*** *of windsurfers congregate on the water. Most coastal resorts now offer windsurfing to vacation-makers as a recreational watersport.*

manoeuvrable boards equipped with foot straps for better wave riding and jumping. Today, there is a professional racing circuit generating millions of dollars of income for professionals around the world.

Types of windsurfing

There are three basic types of windsurfing – longboard, slalom and wave sailing. The longboards are the most versatile boards, performing well in most wind and sea conditions. Shortboard or wave sailing requires greater skill and more wind since the boards are smaller and less versatile, designed to work well in planing mode at wind speeds of 13 to 50 miles (21 to 83 kilometers) per hour. Slalom sailing involves speed and high-speed maneuvers, requiring winds of at least 12 miles (19 kilometers) per hour.

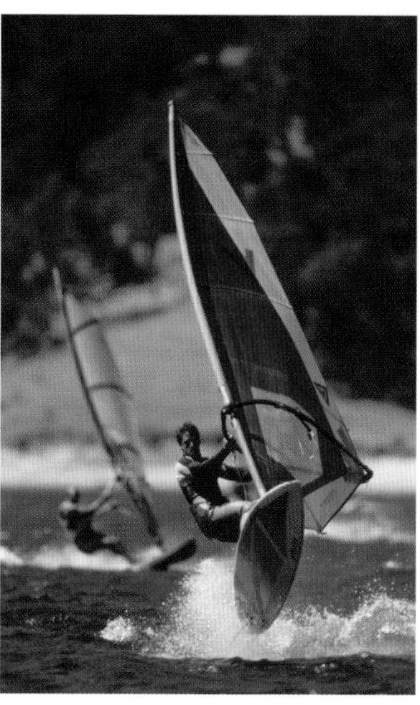

▲ ***A windsurfer*** *on Lopez Lake, San Luis Obispo County, California, USA, uses the wind to lift the board out of the water. The surfer holds onto the boom, controlling the sail for direction and speed.*

Windsurfing equipment

Basic windsurfing equipment consists of a board with a fin, or fin centerboard combination, at the bottom and rig attached to the top by means of a universal joint. The fin or centerboard is like the keel of a boat keeping the craft going forward rather than sideways, while the rig includes the sail that catches the wind, providing forward momentum to the craft. When going slowly, the centerboard should be down and the mast forward, while at higher speeds, the mast is run to the back of the track and the centerboard is retracted. The rig includes a mast or pole that holds the sail up and the boom, the part that the windsurfer hangs on to, while the universal joint allows the rig to swivel and tilt independently of the board. This permits the windsurfer to steer by simply angling the sail one way or the other.

As in the case of surfboards, windsurfing boards are of three types – kneesurfers, which are slow and as their name suggests are sailed kneeling down, longboards and shortboards. The most versatile board, the longboard, is good for most wind and water conditions and features a centerboard, a foot-adjustable mast track and footstraps. The shortboard is the fastest and most manoeuvrable type of board and combines simplicity of design with high wind performance compared to the longboard. It features a basic mast track, three or four footstraps and a high performance fin.

Added to the diversity of boards are a variety of sail types that range from light wind sails, which are light and without battens, through slalom sails with battens, to race sails. The latter are the most rigid and stable, with camber inducers and up to seven battens; these sails are heavy and relatively difficult to rig.

The first divers

Man's physical penetration beneath the surface of the ocean in order to observe at first hand, is constrained by the fact that we breathe air. Free diving was, and still is, practiced in many parts of the world to collect valuable resources, such as pearls in the Persian Gulf or Japan, or for underwater spear fishing, but such dives are inevitably rather shallow and of short duration.

One of the earliest recorded diving operations was that of the Greek Scyllis, who, together with his daughter Cyane, succeeded in cutting the anchor cables of the Persian fleet immediately before the Battle of Salamis – a feat which was achieved by swimming underwater, using a snorkel of hollow reed. In 1690, Edmund Halley designed a diving bell, which received its air supply through a leather pipe leading from a weighted, air-filled barrel. As the air supply in one barrel was exhausted, another was lowered down. By the end of the 18th century, diving suits had been developed which were fed with air pumped into the suit from the surface.

The invention of the diving helmet led to the development of the Siebe open-diving suit in 1819. The suit was used in salvage work on the *Royal George* in the 1830s. Later, in 1837,

◀ ***A man stands** with an enclosed diving suit, 1918. Air was pumped into the suit allowing the diver to walk along the bottom of the ocean. Such suits were used in salvage operations on sunken ships. Later developments in breathing apparatus removed the need for the all-in-one suit and allowed for much greater flexibility.*

Siebe produced the first enclosed diving suit to incoporate a diving helmet. This remained the standard diving equipment throughout the world for nearly 100 years.

Scuba development

In 1943 Jacques Cousteau and Emile Gagnan carried out the first successful tests of the Self-Contained Underwater Breathing Apparatus, the now well-known Scuba equipment. This represented a major breakthrough in underwater exploration, freeing the diver from the cumbersome diving suit and airlines, which restricted the range of movement under the ocean surface.

A scuba diver can move freely, carrying an air supply in canisters on a backpack which is breathed through a regulator. Initial trials involved the use of tanks of pure oxygen, but below 25 feet (8 meters), pure oxygen is toxic. Air is normally used to depths of around 200 feet (60 meters), below which nitrogen, which makes up around 80 percent of air, becomes narcotic. At greater depths, a mixture of oxygen and helium can be used. Since a diver breathes air or a mixture of gases at the same pressure as the surrounding water, nitrogen or helium are absorbed into the bloodstream. If the diver ascends too rapidly, there is not enough time for the gas to diffuse out through the lung surface. Bubbles may form in the blood stream, causing the painful condition known as the bends; in extreme cases, this can lead to unconsciousness or even death. The deeper the dive and the longer the period of time spent at a particular depth, the greater the amount of time needed for decompression. Divers working at great depth are frequently saturated with inert gas prior to the dive and rest in pressurized chambers between spells of work, so that only one lengthy period of decompression is required following completion of the job.

Modern scuba diving can be enjoyed by most individuals and one does not need to be an Olympic swimmer to become a scuba diver since the basic requirement is to be able to swim just 200 meters

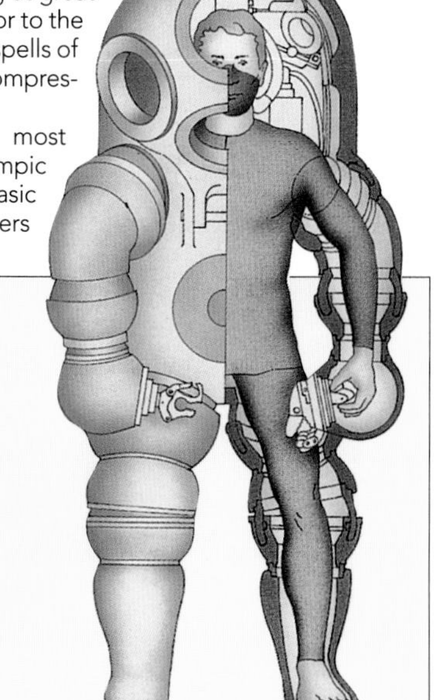

JIM

The atmospheric diving suit, known as JIM after its inventor, is a compact, robotlike suit with tight presure seals and specially designed articulated joints which allow the diver to flex his arms and legs, and work at depth.

Modern JIMs are made of carbon fiber, reinforced plastic and aluminum alloy. Using such materials, most JIMs are capable of diving down as far as 1500 feet (450 meters). Some JIMs have thrusters allowing the diver to remain in position and work in midwater. Since the diver is always at atmospheric pressure, there is no need for lengthy decompression.

using any stroke. Modern scuba diving attracts millions of participants, with specialist resorts in exotic locations offering diving experiences in freshwater and marine environments: on coral reefs and wrecks, in caverns, lakes and rivers. Part of the reason for the growing appeal of scuba diving has been the development of comparatively inexpensive equipment and technological improvements that result in increased diver safety. Improvements include buoyancy vests for flotation, wetsuits for thermal insulation, depth gauges and effective regulators. In addition, there is the easy accessibility of dive tanks through the numerous dive operators worldwide.

Diving equipment

A scuba diver carries with them their own supply of compressed air in a cylinder made either of aluminum or steel and ranging in volume from 1416 to 2679 cubic liters (50 to 94.6 cubic feet). The working pressure to which a tank is filled ranges from 122 to more than 272 atmospheres (1800 to 4000 pounds per square inch). Aluminum tanks have advantages over steel: they do not corrode in the same way as steel cylinders, since the oxide that forms creates a protective surface that prevents further corrosion, and the tanks come with a flat bottom, allowing them to be stood upright. Disadvantages include seizing of the valve in the cylinder neck, which rarely happens with steel tanks, and being more easily dented and damaged than steel tanks.

▼ ***The development*** *of diving apparatus has allowed the diver independence from surface-supplied air. H.A. Fleuss developed the first practicable self-contained diving apparatus. It was able to filter out carbon dioxide from the exhaled air and replace it automatically with the equivalent amount of oxygen.*

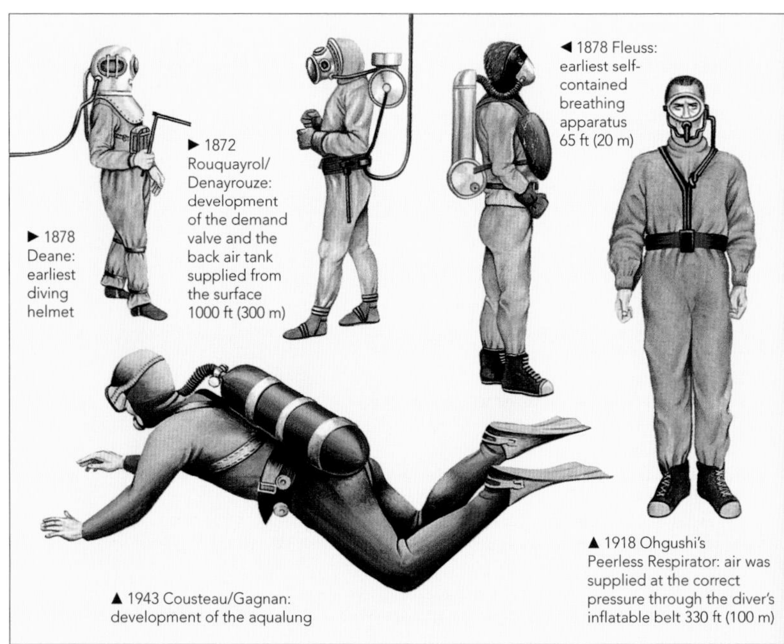

► 1878 Deane: earliest diving helmet
► 1872 Rouquayrol/ Denayrouze: development of the demand valve and the back air tank supplied from the surface 1000 ft (300 m)
◄ 1878 Fleuss: earliest self-contained breathing apparatus 65 ft (20 m)
▲ 1918 Ohgushi's Peerless Respirator: air was supplied at the correct pressure through the diver's inflatable belt 330 ft (100 m)
▲ 1943 Cousteau/Gagnan: development of the aqualung

► ***A scuba diver*** *explores a coral reef in the Red Sea. A shoal of sea goldie* (Pseudanthias squamipinnis) *or scalefin anthias fish swim out to greet the diver. Sea goldie feed on zooplankton. They are territorial, with males living in haremic groups with a number of females. The male grows up to 6 in (15 cm) in length, compared to the 3 in (7 cm) female.*

Between the tank and the diver's lungs are two important pieces of equipment. The first is the tank valve that contains a metal 'burst disk' to prevent the tank from being overfilled. The simplest, the K valve, simply turns on or off, while the J valve maintains a reserve of air in the tank to permit a normal ascent. The submersible pressure gauge tells the diver what pressure remains in his tank during a dive. Another vital piece of equipment is the regulator, which reduces the pressure of the air inside the tank to a breathable level in two stages. The first, which attaches to the tank valve, reduces the pressure to about 9.3 atmospheres (140 pounds per square inch) and this connects via a hose to the second stage which reduces the pressure further, to that of the surroundings.

A swimmer normally floats. In order to dive, they must expend energy, but weights are used to adjust the individual to neutral buoyancy during a dive, such that they neither float nor sink. Fins increase the efficiency of swimming and diving, and improve the speed of movement underwater. Modern masks can be modified to accommodate lenses for those who normally wear spectacles. Positive buoyancy is important when a diver needs to rest at the surface to lift items collected during a dive, so a buoyancy vest with an oral inflation tube becomes an invaluable aid. Buoyancy compensators help the diver control their weight throughout a dive. Inflation at the surface increases buoyancy, deflation reduces buoyancy during a descent, and air can be added underwater to achieve neutral buoyancy. Most buoyancy compensators are designed with a backpack that holds the scuba cylinder.

Wet and dry suits

One of the problems faced by all mammals in water is a more rapid loss of body heat from the surface than when on land, and man is no exception. Some kind of exposure protection when diving not only assists in retaining body heat but also protects against bangs and scrapes. The two basic types of suit are the wet suit and the dry suit. Generally, the colder the water, the thicker the suit and more extensive its coverage. Wet suits allow water inside and are made from foam neoprene or a special nylon material (Spandex); the better the fit, the less water inside and the warmer the suit.

Wet suits are good for water temperatures ranging from 60° to 85°F (15.6° to 29.4°C), while in colder waters a dry suit is advisable. As its name suggests, this suit type excludes water and is generally worn with undergarments. Dry suits are designed not to compress at depth, so the thickness of

▼ ***Diving sites*** *in the Red Sea. The extensive coral reefs, marine life, and shipwrecks make the Red Sea one of the world's most popular places to dive.*

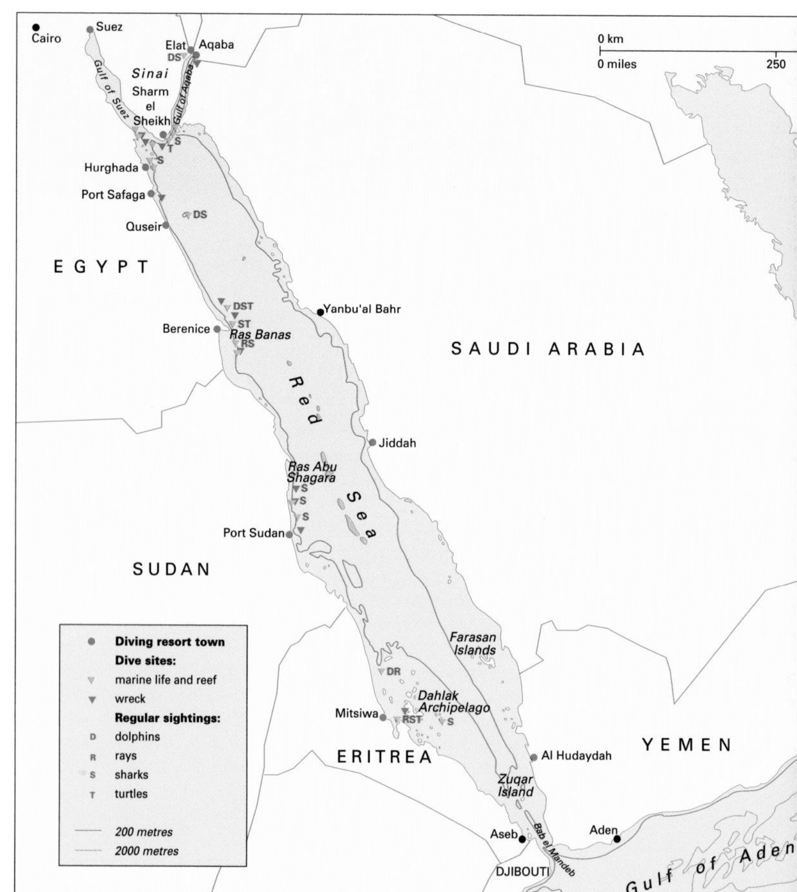

the insulation remains fairly constant. Diving suits come with accessories, including boots and gloves or mitts. Hoods are an important way of retaining the body's warmth, and can reduce heat loss from 20 to 50 percent depending upon the water temperature. Separate hoods generally have a skirt that ends at the base of the neck. Water circulates through this joint, and cold water divers generally prefer attached hoods.

Underwater photography

Along with the advances in diving equipment have come technological changes that allow photographs to be taken underwater. Professional photographers use expensive waterproof housings for large cameras and their associated flash systems, but small, inexpensive underwater cameras can be used by any diver. The consequence has been an explosion of images, easily available in books and film and on the web, such that our knowledge and understanding of life beneath the sea is now considerably greater than it was just 30 years ago.

▲ ***A diver films*** *a coral reef in the Bahamas with a specially designed underwater video camera. Advances in underwater technology have given nature photographers access to a world that was previously unfilmed.*

Limitations

Despite these improvements in equipment, which make scuba diving easier, safer and more fun than in the past, there is still a limit to the depth and length of time that divers can remain submerged. The time limits are determined by the amount of nitrogen dissolved in the body, and the deeper the dive, the shorter the time. At 130 feet (40 meters), the time limit is usually set at five minutes. At shallower depths such as 50 feet (15 meters), the maximum time is somewhere between 70 and 80 minutes. Dive tables were originally developed by the navy to allow calculation of the required decompression times and safety limits for dives at different depths. The complexity of these calculations has now been removed with the development of dive computers worn like a watch that tell the diver how much time can be spent at different depths and how much time should be spent during the ascent at different depths to allow outgassing of the nitrogen dissolved in the blood.

The use of submersible craft is necessary at greater depths than those which can be reached by divers, and they have an added advantage in that the operators do not need to undergo extensive periods of decompression. Nevertheless, skilled divers are still needed, since many of the tasks required in the maintenance of oil rigs and underwater structures could not be achieved without the manual dexterity of a human operative.

Game fishing is the pursuit, with rod and line tackle, of a wide range of large fish including billfish, sailfish, wahoo, tarpon and the larger species of active sharks, such as the mako and tiger shark. Most game fishing is done from charter boats, and the species sought may depend both on the preference of the angler and what is available in the area at that time of year.

Methods

Game fish are caught with methods that include lure trolling, dead-bait trolling and live-bait trolling. Since many different species of fish are caught by trolling, it is often difficult to predict what will be caught, but in deep open water, species such as the marlin, sailfish and tuna are the regular catch. Big game boats typically fish four rods at a time, one from each side of the boat held clear on outriggers, and two from the stern. Each rod has a different length of line played out and usually uses a different lure, some submerged and some that skip along the surface. The combination of different lures at different distances from the boat is designed to create sufficient disturbance to suggest to the predatory fish below that a shoal of fish is at the surface. Sometimes a teaser is drawn behind the boat, designed to provide maximum flash and movement at the surface and attract the fish toward the lures.

▼ ***Deep-sea fisherman*** *catch a marlin. In 1950 US novelist Ernest Hemingway started an international marlin tournament at Havana, Cuba. In 1960, Cuban leader Fidel Castro won the competition. Hemingway referred to the marlin rich waters of the Gulf Stream as the 'Great Blue River'.*

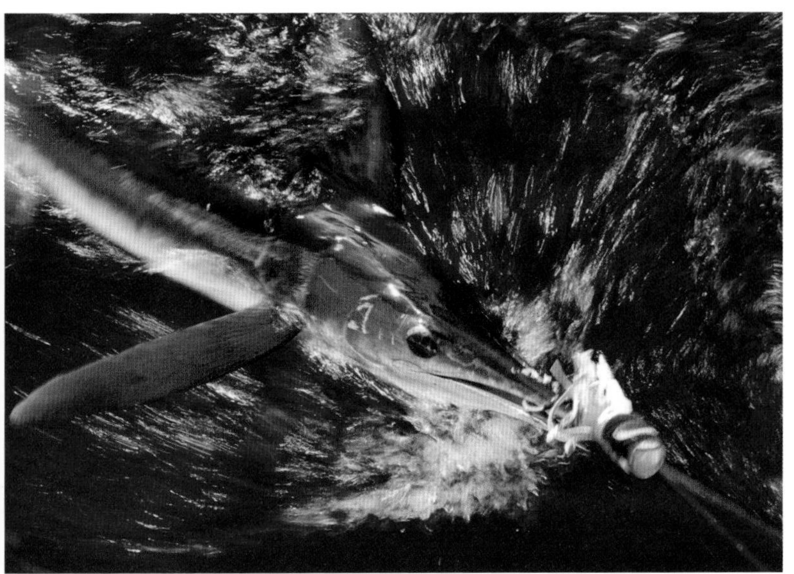

▲ ***Sport fishermen*** *hawl a white marlin* (Makaira albida) *onto their boat. The speed and size of marlin make them a popular gamefish. Black and blue marlin weighing more than 1400 lbs (635 kg) have been landed. They swim at speeds up to 50 mph (80 km/h). The blue marlin is the largest of the billfish.*

Bait

The appearance and smell of a dead bait makes it potentially more attractive to predatory fish than an artificial lure, but if the bait is not made to move through the water in as natural a manner as possible, then that advantage is lost. Dead bait should be rigged on the hook so that it moves forward and should be trolled at a speed that ensures it 'swims' naturally. To achieve this, smaller dead baits have the hook placed inside the fish with the line passing through the mouth, which is then tied shut to prevent water resistance from moving the bait in an unnatural manner. Large baits need to be trolled at much slower speeds than small baits and lures. Live bait trolling also needs to be done slowly so that the bait does not drown, and larger baits such as bonito and dolphin fish are generally tied to the hook using a bridle rig rather than being impaled.

Live bait drifting is another technique that can be used in the vicinity of pinnacles, reefs or wrecks in shallower water and can result in some fine sport with actively fighting fish such as amberjack and bonito. Sharks, such as the mako and blue sharks, are mainly open-water species which feed on pelagic shoal fish, such as mackerel, and they can be attracted to the vicinity of the bait by a trail of 'rubby dubby' in the technique known as chumming. 'Rubby dubby' consists of a mixture of oily fish, such as mackerel, bran and fish oil, pounded so that the mixture is composed of different sized pieces. A mesh bag full of the mixture, when hung from the stern of the boat, leaves an oily slick of particles of different sizes, which sink to different depths, providing a three-dimensional scent trail toward the boat.

The origins of sailing

The origins of sailing undoubtedly lie in the first attempts at maritime travel and, based on the different types of sails, boat designs and methods of navigation, probably originated independently in different parts of the world. Early European sailing ships, the square-riggers, were built to go downwind; the sails were rectangular in shape, and set ahead of the masts at right angles to the length of the ship and supported at the top by yards. Most square-rigged ships could sail across the wind and at all angles downwind, but even with sails hung close to the center line of the boat, square-rigged ships did not sail well into the wind. The 15th-century Spanish and Portuguese ships were all square-rigged. As the early European explorers opened up the world trade routes, they followed established weather systems that enabled the ships to complete their voyages by sailing downwind.

The Polynesians developed multihulled craft for their ocean voyages, with large triangular sails supported by a mast and side spar. The Arabs regularly sailed their dhows across the Indian Ocean, and in the Mediterranean other craft, such as feluccas, were developed. Many of these craft had the sails rigged fore and aft rather than at right angles to the length of the boat, an innovation that, enabled them to sail upwind more easily than the square-rigged vessels. Sails were rigged ahead of the mast with the mainsail set behind the mast and attached by its leading edge (luff) to the mast itself.

Square-rigged ships continued to be constructed well into the 20th century, and included the famous tea clippers

▼ ***A dhow sails** into the harbor at Stone Town, Zanzibar. Dhows are one of the earliest forms of sailboats. The dhow's use of a lateen rig was a development that enabled boats to sail close to the wind.*

and commercial sailing ships that carried cargoes of industrial goods, wool, grain and tea to and from Europe, Australia and the Far East. Although suitable for long-distance sailing, such vessels were less suitable for inshore purposes, so many fishing vessels, pilot vessels and coastguard cutters were rigged fore and aft, and it was from these working vessels that most of the early pleasure yachts were developed.

Early fore- and aft-rigged boats were often gaff-rigged, with the main sail being four-sided and having the top supported by a spar or gaff. Modern fore- and aft-rigged boats are generally based on the Bermuda or Marconi rig, familiar as the triangular mainsail attached at the leading edge to the mast and with the lower edge attached to a boom. The Bermuda rig has the advantage of being able to point closer to the wind than the gaff rig, but the disadvantage of requiring a longer mast.

▲ ***Croatian skipper*** *Frano Brate heads to the top mark during Day Four of the ACI HTmobile Cup in Split, Croatia, May 30, 2003. The Cup, the most famous regatta in Croatia, is one of ten match-racing events comprising the Swedish Match Tour, the most prestigious international match-racing series. Crews compete on standardized boats.*

Sailing for pleasure

As early as the 1800s, fishing fleets and pilot boats often organized annual race days for their working craft. The term yacht, which has become synonymous with sailing for pleasure, appears to be derived from the Dutch word, *jaght* (or *jacht*) and a *jachtschip*, or 'ship for chasing', which was a swift, light vessel of war. During his exile, Prince Charles, later King Charles II of England, spent some years in Holland and on his return was presented with a yacht the *Mary* by the Dutch, heralding the introduction of yachting into England.

Sailing clubs started to appear in the 18th century. In 1775 the Cumberland Fleet was founded in England and organized racing on the River Thames. In Ireland, the Cork Harbour Water Club, now the Royal Cork Yacht Club, was founded as early as 1720. While yachting appears to have originated in Europe, many developments occurred along the east coast of America where some of the fastest clipper ships of the mid 19th century were designed and built by Donald McKay of Boston. The Americans favored schooners of shallower draft and broader beam, than the more narrow, deep drafted craft developed by the British. The New York Yacht Club was founded in 1844. In 1851 a syndicate of American yachtsman sailed the schooner *America* to England where she won the '100 Guinea Cup', as it was known, in a race around the Isle of Wight. She won by such a margin that no European yacht would accept the challenge of sailing against her, and the cup returned to America where it became famous as the America's Cup. The slim bow of the *America* also initiated a more trim design for yachts. The sport was also becoming worldwide: the Royal Sydney Yacht Squadron was founded in Australia in 1862.

Many of the early one-design yachts were small half-decked keelboats sailed by a crew of two or three while others were even smaller – the dinghies. Both the Star Class, designed in 1911 and raced in the 1932 Olympics, and the Snipe Class, designed in 1931, remain international class dinghies in widespread use. The development of these craft opened the sport to a greater number of people. After World War II, yachting increased dramatically with demand spurred by specialist magazines and supply fulfilled by the introduction of cheaper, mass produced fiberglass boats. By 1993, the sales of Laser Class dinghies, designed in 1971, had topped 150,000.

Yacht racing

Racing appears to have been well established by the mid 1800s, but a perennial problem was that races tended to be won by the bigger boats since the maximum speed of a displacement hull is related to about 1.4 multiplied by the square root of the waterline length. This is somewhat simplistic, since other factors such as beam, total displacement and sail area all affect the maximum speed, but the dominant factor is the length of the waterline. This led to the development of handicapping or rating craft, thus enabling craft of dissimilar design to be raced together. The rating rules, however, were never perfect and the situation became complicated in that different clubs operated under different rules and very different rating standards were developed on each side of the Atlantic. Only in 1970 did the International Offshore Rule (IOR) come into widespread use for international competition. The IOR has itself been replaced by the International Measurement System (IMS), while racing under the Performance Handicap Racing Fleet system is popular in some parts of the world.

▼ ***The start of a yacht race*** *The yachts are all of a similar design to ensure fair racing. The competition is based on the skill of the crew aboard the yacht rather than the power of the yacht itself.*

To avoid the problems associated with racing under different rating or handicap systems during the 20th century, several one-design classes of yacht were developed specifically for racing. Generally at the lower end of the size range, crews of such craft compete on equal terms and, in theory at least, the best crew wins. The crews of all levels of competition are open to women.

The courses for standard yacht racing have no standard design and can vary greatly. The course for the Olympic competitions was devised as an excel-

▲ **Matador *racing yacht*** *with crew. At 85 feet (26 meters), Matador was the largest International Offshore Rules (IOR) yacht ever built. She remained undefeated in a total of 52 races, eight regattas and two World Maxi Championships.*

lent test of skill, and this design is commonly used. The yachts race around a triangle marked by buoys, and a competition is normally made up of several races. The final winner is the crew who accumulate the fewest total points; the victorious crew of each race receives no points, while second place is awarded 3 points and third place 5.7 points.

The Planing dinghy

A major breakthrough in dinghy racing came with the design of the planing dinghy which, unlike the displacement keel-boats, rise in the water like a speedboat, thereby reducing their displacement and the surface in contact with the water. The first American scows, rather shallow boats with rounded bows, planed, while in South Africa classes of 25- and 20-foot scows based on the American designs, also planed at high speeds. The designer Uffa Fox is generally credited with being the father of the planing dinghy, developing an interntational 14-foot class dinghy, *the Avenger,* in the 1930s. In her first season, *the Avenger*, won 52 of the 57 races she entered. This revolutionary design forms the basis for most of today's high-performance sailing dinghies. Combined with modern lightweight materials, modern dinghies can plane much faster at lower wind speeds than in the past.

Ocean voyaging and racing

The first transatlantic yacht race occurred in 1866 when three US schooners crossed from New Jersey to England. The first known single-handed round-the-world yacht voyage was

◀ **Endeavour**, *Thomas Sopwith's British J-Class America's Cup challenger in 1934, now fully restored, seen sailing off the coast of San Diego, USA.*

made by Captain Joshua Slocum in 1895 in his 10.9-m (36-foot) sloop, *the Spray,* which was unique for her time in having self-steering gear. Sailing from Boston, he returned three years and two months later having sailed outward through the Straits of Magellan since the Panama Canal was not open at that time. Modern ocean voyagers take this short cut from the Atlantic to the Pacific and avoid the more perilous southern route round Cape Horn.

Offshore and ocean racing has also developed apace since the Fastnet Race, from Cowes to Plymouth via Fastnet Rock off the coast of Ireland, was first sailed in 1925 – the year the Royal Ocean Racing Club was founded in England. Even before that, in 1906, the Bermuda Race, from Newport, Rhode Island to Bermuda in the North Atlantic was inaugurated although it lapsed some years later. Both races are now sailed biennially. The Sydney to Hobart Classic, initiated in 1945, is sailed annually.

Longer races, including the Transpac from Los Angeles to Hawaii; the Cape to Rio, from South Africa to Brazil; and the Transatlantic Single-Handed Race from England to the USA, have all become well established over the last fifty years. Two of the more spectacular races are the Volvo (previously the Whitbread) Round-the-World Race for full-crewed yachts and the Around Alone, single-handed round-the-world race. These races take the yachts deep into the Southern Ocean, while the BT Challenge tests sailors by forcing them to sail around the world against, rather than with, the wind.

The ruling body for the sport is the International Sailing Federation (ISAF), formerly the International Yacht Racing Union (IYRU).

Mono-hulls and multihulls

Over the last 50 years, design of mono-hulled craft has been revolutionized as boats became lighter and flatter. Fin keels have now replaced the traditional wineglass-shaped keel built into the hull sections of older boats. Rudders are separate from the keel, being set at the aft end of the craft, and wood has given way to fiberglass, Kevlar, and Kevlar sewn carbon-fiber. Modern ocean-going racing yachts are now lighter, faster craft, and these materials have become incorporated into other types of yachts as well.

In contrast to the mono-hulled craft, developed and refined in the Atlantic by the Europeans and Americans, multihulled craft were constructed by the early Pacific Islanders for ocean-going voyages more than 1000 years ago. Introduced into western yachting comparatively recently, multihulled craft were not initially very popular, but with the introduction of new high-tech materials, they have become extremely fast and more popular as a result, and perhaps the best known multihulled classes are the 16-foot and 14-foot Hobie Cats.

AMERICA'S CUP WINNERS (1962–2003)

Year	Winning boat	Country	Losing boat	Country	Score	Winning skipper
1962	*Weatherly*	US	*Gretel*	Australia	4-1	Emil Mosbacher, Jr.
1964	*Constellation*	US	*Sovereign*	England	4-0	Bob Bavier, Jr
1967	*Intrepid*	US	*Dame Pattie*	Australia	4-0	Emil Mosbacher, Jr.
1970	*Intrepid*	US	*Gretel II*	Australia	4-1	Bill Ficker
1974	*Courageous*	US	*Southern Cross*	Australia	4-0	Ted Hood
1977	*Courageous*	US	*Australia*	Australia	4-0	Ted Turner
1980	*Freedom*	US	*Australia*	Australia	4-1	Dennis Conner
1983	*Australia* II	Australia	*Liberty*	US	4-3	John Bertrand
1987	*Stars & Stripes*	US	*Kookaburra III*	Australia	4-0	Dennis Conner
1988	*Stars & Stripes*	US	*New Zealand*	N. Zealand	2-0	Dennis Conner
1992	*America 3*	US	*Il Moro di Venezia*	Italy	4-1	Bill Koch
1995	*Black Magic*	N. Zealand	*Young America*	US	5-0	Russell Coutts
2000	*Team N. Zealand*	N. Zealand	*Luna Rossa*	Italy	5-0	Russell Coutts
2003	*Alinghi*	Switzerland	*Team N. Zealand*	N. Zealand	5-0	Russell Coutts

▲ ***A 1930s poster** for the cruise line 'Italia'. In 1933 their luxury liner* Rex *became the only Italian ship to be given the Blue Riband for the fastest crossing of the Atlantic.*

In the era before commercial airlines, traveling for business or pleasure was undertaken by ship and train, and often involved lengthy periods confined in small cramped quarters, periods of inevitable seasickness, and a rather monotonous diet. Ocean passenger lines plied between America and Europe, between Europe and the Far East including Australia and New Zealand, and to more exotic destinations in Africa, India and Latin America. With the advent of steamships, the length of passage to America from Europe was reduced, and many passenger lines competed for customers by providing luxury accommodation for those who could afford to pay.

As taking vacations became more common, the idea of an ocean cruise became the dream vacation for many people in the 1940s and 1950s. Ships were constructed with stabilizers that reduced the effect of the boat's rolling and pitching motion. These ships often had well-appointed cabins, dining rooms and ballrooms, deck-side swimming pools, and games such as deck quoits. Cruise-line operators developed programs in exotic areas with long periods at sea, broken by port calls where the adventurous could visit for a short period, being taken ashore by small boat in some of the more remote locations, such as the islands of the Pacific.

Ocean yachts and liners still offer cruises of varying length in many parts of the world, but they are less in demand than before, since the opportunities for visiting faraway places have been vastly expanded by modern jet travel. In addition ocean cruising by yacht has become a real possibility for many people since the costs of an ocean-going craft have decreased and *per capita* income in Europe and America has been rising over the last forty years.

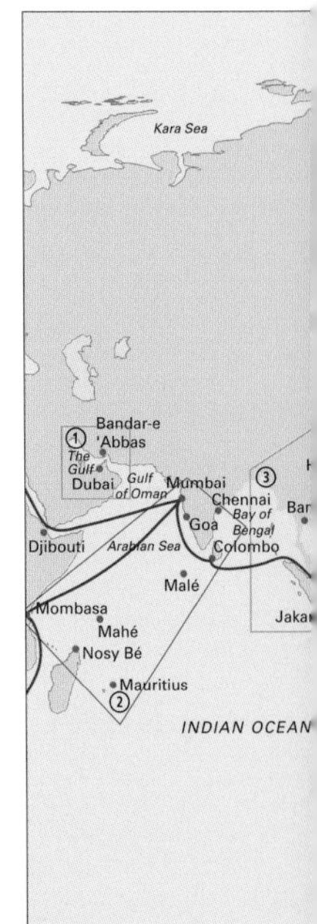

Long-distance cruising

For many people who sail offshore boats at the weekends, their ultimate dream is to undertake an ocean voyage, and

For those who live in colder climates, the lure of the tropics is irresistible. Just as ocean liners have become more comfortable than their predecessors of a century ago, so the modern cruising yachts have been designed for both optimum sailing performance and the comfort of the crew. Modern navigational and other electronic devices make sailing and communication easier.

The lure of ocean cruising was greatly enhanced by the cruises and writings of Eric and Susan Hiscock, who were perhaps the world's most famous cruising couple. They started cruising in 1950, sailing from England to the Azores and back in the 7-m (24-feet) *Wanderer II*. In 1952 the Hiscocks commenced their first circumnavigation aboard the 9.1-m (30-foot) *Wanderer III*, and Eric started producing well-written and beautifully illustrated magazine articles and books that made the '*Wanderers*' (they owned five successive vessels) and their owners well known around the world.

The number of individually owned yachts cruising in the Caribbean, the Mediterranean and even the Indian and Pacific Oceans has grown exponentially, such that marinas are a common sight in many coastal areas where boats may be kept and maintained when not in use by their owners. Small-yacht cruises are also available in many tropical coastal regions catering for ship-based diving vacations, whale watching, or big game fishing, or just plain sailing, according to the wishes of the charter.

▼ ***Map showing*** *main cruising areas and the principal round-the-world cruise routes. The major ports are also shown.*

① The Gulf cruise area
② Africa–India cruise area
③ Far East cruise area
④ Australasia/South Pacific cruise area
⑤ Hawaiian Islands cruise area
⑥ Alaska/Canada cruise area
⑦ Mexican Riviera cruise area
⑧ Caribbean/Florida cruise area
⑨ South America cruise area
⑩ US East Coast cruise area
⑪ West Africa/Atlantic islands cruise area
⑫ Norwegian Fjords cruise area
⑬ Baltic cruise area
⑭ Mediterranean cruise area
⑮ Black Sea cruise area
▬ Round-the-world route

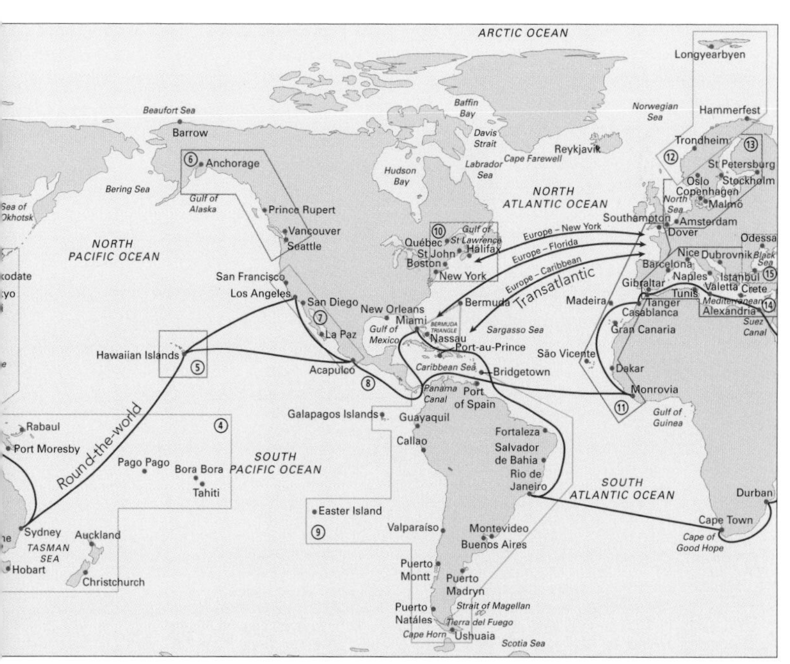

The growth in international tourism since World War II has been staggering, reaching nearly 700,000 tourists in 2000 and generating a total of US$463 billion in revenue. Prior to World War II, it was only the wealthy who traveled, since the costs, both in money and time, were too great for the average worker. Even so, tourism generally involved only comparatively short-distance travel within the country of origin or to neighboring countries. The recent growth in international tourism reflects numerous social and economic changes, including greater personal income and cheaper and faster travel, such that reaching tropical destinations in southeast Asia takes only 11 hours from Europe.

In the early 20th century, day trips to the seaside by bus or train characterized British tourism, with the workers from the wool and cotton towns of Yorkshire and Lancashire visiting local resorts, such as Scarborough and Blackpool, and Londoners visiting Brighton. After the war, vacation camps sprang up, pioneered by Billy Butlin who bought old army huts and erected them in coastal locations providing a range of entertainment and activities on site, and bringing the costs of a stay at the seaside within the reach of the majority of the population.

The late 1950s and 1960s saw a growth in European tourism with increasing numbers of visitors from the colder northern countries moving south to the Mediterranean during the summer to enjoy seafood, sunshine and the beaches. Even now some 100 million tourists a year visit the Mediterranean, with other visitors going to more exotic locations including Turkey, Algeria, Israel and Egypt. In 2000, half of the world tourism receipts were collected in Europe and around 18 percent in the Americas.

▼ **Bridlington, It's Quicker by Rail** *by Andrew Johnson. This London and North Eastern Railroad poster dates from 1928. The advent of the railroad greatly increased the number of visitors to seaside resorts. Tourists and day-trippers flocked to the golden sands at Bridlington, Humberside, northeast England.*

► ***Sun-seekers*** *crowd the beach on the Bay of Biscay, Biarritz, south-west France. Biarritz has been a fashionable seaside resort since 1854, when Napoleon III's wife Eugénie built a villa there. Visited by Queen Victoria, Edward VII, and Alfonso XIII of Spain, Biarritz began to call itself 'the queen of resorts and the resort of kings.' The town prospers on the tourist industry and its population quadruples in the summer months.*

Package tourism

The model of a self-contained resort providing all manner of sport and entertainment was copied both in Europe and elsewhere and ushered in the era of mass tourism, with individuals buying a 'package', an all-inclusive arrangement paid for in the country of origin and covering travel, accommodation and meals. Coastal resorts now cater for all types of marine sports, including snorkeling, scuba diving, windsurfing, yachting and canoeing, while at the same time providing a range of evening entertainment, from bars and discos to more exotic shows and revues.

Tourism in the Maldives, for example, started in the late 1960s, with the development of self-contained resorts occupying an entire small island and each having a daily program of activities, including boat trips and watersports. Resorts were developed by German, French and Italian tour operators who chartered aircraft to bring a weekly plane-load o tourists, returning with those who had stayed a week or twc weeks. Such resorts, although they provide an exotic experience, insulate the tourist from the difficulties of communication in a strange country with a different language, food anc culture.

The initial attraction of such coastal resorts was their warm water, fresh breezes and the opportunity to try out watersports, in particular scuba diving in exotic environments such as coral reefs. In the 1970s, most resorts began operating their own dive centers, providing rented gear and training for the novice diver, and similar opportunities are now available for windsurfing. Despite the undoubted importance o tourism to the economy of island countries such as the Maldives and Fiji, and the countries of southeast Asia, the financia returns from package tourism are proportionally quite small since the tour operator in the country of origin buys the accommodation, charters the planes, contracts out all the on-site services, and hence makes a greater profit than the individual on-site service providers.

▼ ***Household rubbish*** *litters a sandy beach on the Venezuelan island chain of Los Roques. Pollution like this can have a terrible effect on a local ecozystem.*

Environmental impacts

A further disadvantage of package tourism for the host country is the environmental damage that results from large numbers of people being packed together in a small area. Coastal habitats are particularly sensitive to disturbance, and the impacts of resort construction are often considerable. Building the resort hote and landscaping the grounds results ir increased sediments that smother cora reefs and interfere with tidal and othe currents, causing erosion and loss of the very beaches which provided the initia attraction to the tourist. Huge quantities of waste are produced, including sewage and organic wastes resulting in eutrophication (excessive nutrients) of coasta waters and the growth of algae. Solic wastes including packaging materials cans and bottles, if not disposed of properly, result in littering, spoiling the esthetic appeal of coral sand beaches.

Construction of a resort hotel represents a considerable capital investment and such complexes are often built toc close to the shore to provide the touris with instant access to the beach and nearby sea. Construction of piers or jetties tc provide easy access to boats, changes the local patterns of currents and this unfortunately often results in beach erosion which in turn prompts the resort develop-

▲ ***A beautiful unspoiled*** *beach on Ko Lanta, Krabi province, southern Thailand. The beach is lined with fishing boats. The glorious sandy beaches of southern Thailand attract backpackers from all around the world. Unfortunately, some beaches, such as those on the island of Phuket, have become popular tourist traps with bars and clubs, and have lost much of their natural beauty.*

er to construct groines to retain the sand on the beach, and seawalls to prevent erosion of the coastal structures. As the investment in coastal areas has risen, so has the need to protect that investment against flooding, storm surges, tsunami and the general erosive action of winds and waves. Shorelines are armored, sea walls constructed, groines extended seawards and beaches replenished with sand. Once started, this cycle of protection and loss continues until some of the islands supporting the older resorts in the Maldives are now completely surrounded by seawalls, and the beach sand has been largely lost from the island.

Recognizing that environmental damage represents a major threat to the long term sustainability of the tourism sector, many governments in developing countries have introduced building codes and regulations for coastal resorts designed to reduce their detrimental environmental impacts. The recent development of ecotourism has also sensitized the tourists themselves to the environmental impacts of their visits and resulted in a more responsible attitude, such that few tourists now buy shells and corals, and most are quite responsible in disposing of their litter. Problems remain, however, as the numbers of ecotourists continue to increase. Many tourist destinations now try to attract individual visitors by providing more opportunities for upmarket and diverse experiences that take advantage of the unique cultural, culinary and environmental opportunities provided in their countries. Nowadays it is not uncommon for young people to take a 'gap year' traveling the world and visiting exotic and out-of-the-way places in search of unspoiled beaches and seeking new experiences, but the lure of sea and sand remain strong attractions to all but a few tourists.

Ocean Life

PLANKTON

Unlike on land, where vegetation dominates and trees grow to enormous size, the primary producers of the ocean are of minute size. The smallest phytoplankton, picoplankton, are less than 2 microns (a hair's width) thick while the largest, called macroplankton, exceed 1/12 inch (2 millimeters) in length. Ocean waters support complex communities of plankton, both plants and animals, which form the basis for the food webs that support the larger, multicelled animals.

Since sunlight does not penetrate beyond 650 feet (200 meters) in the oceans, primary production can only occur in surface waters, where it is generally limited by the low availability of nutrients such as nitrogen, phosphorus and, in some areas, iron. The vast majority of ocean primary production is based on single-celled, floating plants, the phytoplankton. There are two dominant groups: the diatoms, which are characteristic of colder waters, and the dinoflagellates, which dominate the warmer water areas. Although most phytoplankton are unicellular, some grow in chains or form spherical colonies of larger size. Large colonial phytoplankton are characteristic of areas of upwelling, where nutrients are more abundant than in the open ocean.

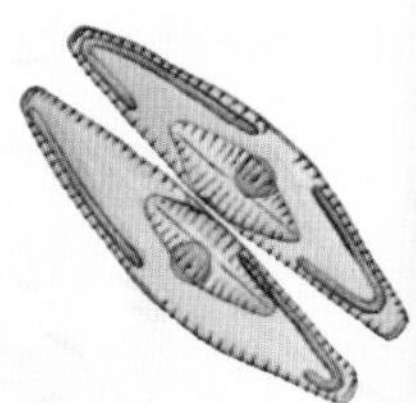

▲ ***Marine diatom*** *Diatoms are single-celled algae.*

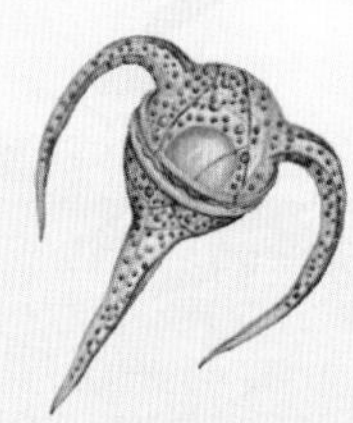

▲ **Dinoflagellates** *are most found in tropical and subtropical waters.*

Starting the chain

Herbivorous planktonic animals eat most of the phytoplankton; the most abundant of these are copepods, small crustaceans whose constantly moving limbs sweep the phytoplankton toward their mouths. The herbivorous plankton possess straining mechanisms for filtering the phytoplankton out of the water; crustaceans sweep food up using hairy, modified legs, and tiny animals such as salps filter the water through their barrel-shaped bodies. Even larger animals, such as the baleen whales and basking sharks, rely on plankton for food. The majority of phytoplankton production of the oceans is grazed by these herbivores, in contrast to communities on land where plants store energy in tissues which are not eaten but decompose under the action of bacteria and fungi. The turnover time

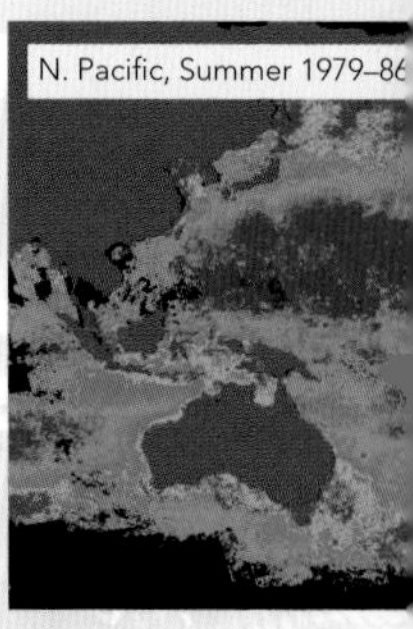

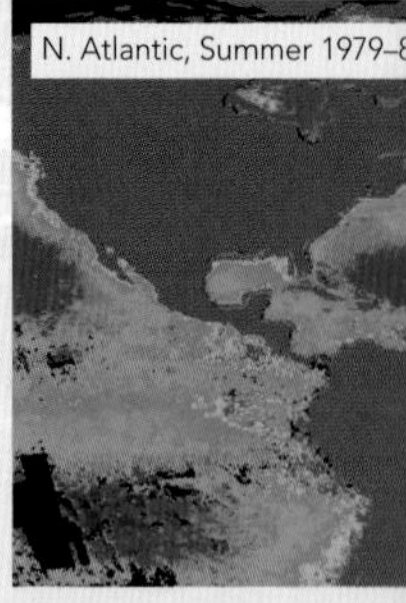

► ***Phytoplankton populations*** *have declined substantially in northern oceans since the early 1980s. The world maps shown here compare satellite data collected during the summer (July–Sept) between 1979 and 1986 and 1997 to 2000, and reflect the changes in phytoplankton concentrations over the last 20 years in the open ocean (away from the coast). Reds show high concentrations, yellow represents medium, green is the next rung down, and blue is low. The largest decline was in the North Pacific where concentrations dropped by 30 percent since the 1980s. In the North Atlantic, concentrations dropped by 14 percent. Scientists from NASA and the NOAA argue that warmer ocean temperatures and low winds may be depriving the tiny ocean plants of necessary nutrients.*

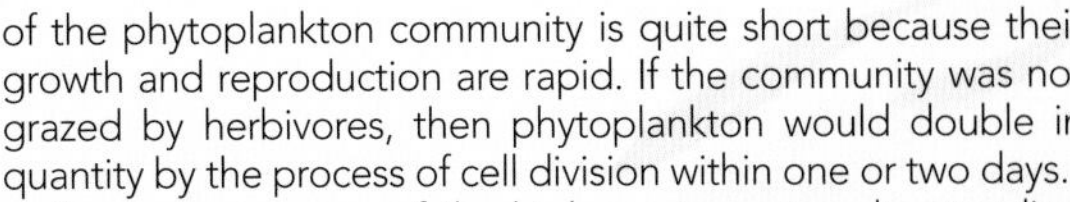

of the phytoplankton community is quite short because their growth and reproduction are rapid. If the community was not grazed by herbivores, then phytoplankton would double in quantity by the process of cell division within one or two days.

As a consequence of the high-grazing rates, the standing stock of phytoplankton in any one area of ocean is normally less than the standing stock of herbivorous zooplankton feeding on them. Although longer lived than the phytoplankton, these zooplankton are also relatively short-lived, surviving for a few weeks and breeding throughout the year in low latitudes, but only during the warmer periods of the year at high latitudes. The planktonic herbivores are preyed on by carnivorous zooplankton, which are in turn fed on by small fishes.

▲ ***Copepods*** *are the dominant herbivores of the plankton community.*

Plankton communities

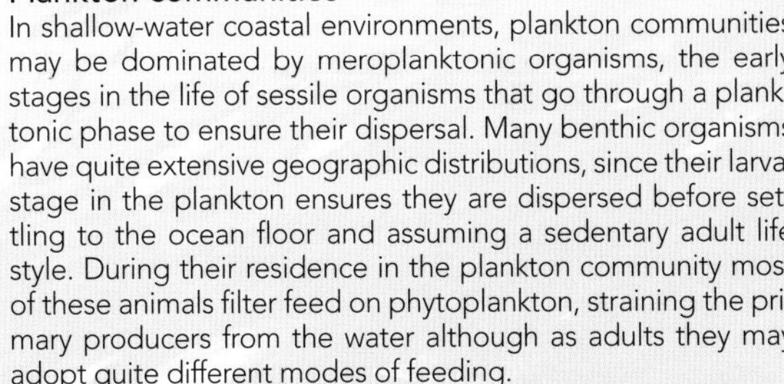

In shallow-water coastal environments, plankton communities may be dominated by meroplanktonic organisms, the early stages in the life of sessile organisms that go through a planktonic phase to ensure their dispersal. Many benthic organisms have quite extensive geographic distributions, since their larval stage in the plankton ensures they are dispersed before settling to the ocean floor and assuming a sedentary adult life style. During their residence in the plankton community most of these animals filter feed on phytoplankton, straining the primary producers from the water although as adults they may adopt quite different modes of feeding.

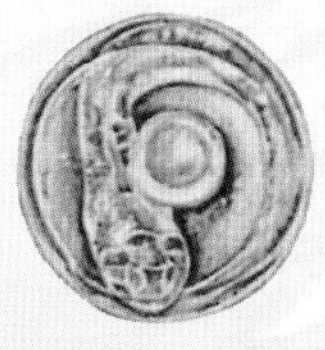

▲ ***Fish eggs*** *are a variety of zooplankton, found worldwide.*

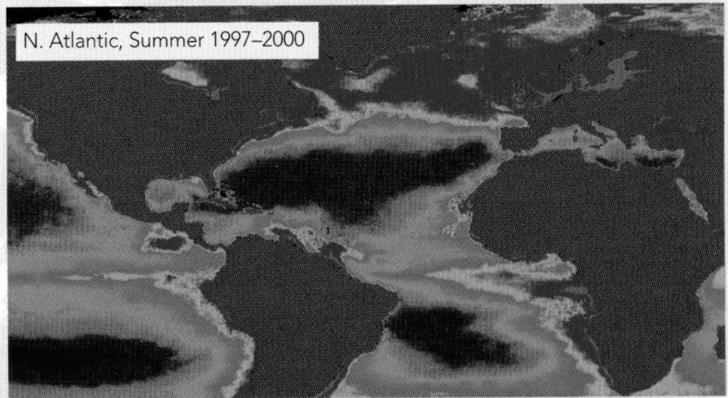

The term 'primary productivity' refers to the conversion of energy from sunlight into the chemical energy of organic molecules through the process of photosynthesis. Photosynthesis only takes place in the presence of sunlight, and with few exceptions, all marine communities depend on the production of phytoplankton for their sources of food energy.

The intensity of sunlight varies over the ocean surface. At high latitudes, intensity is low during the winter and high during summer, resulting in an annual cycle of primary production. Seasonal patterns of productivity vary. In the Arctic, for example, a single peak of primary production occurs in summer during the period of highest light intensity, followed by a peak in zooplankton production. In the North Atlantic, the peak of phytoplankton occurs earlier, in spring, and is followed by the zooplankton peak and a smaller, second peak of phytoplankton during the fall. In tropical systems, the peaks of plankton are far less marked and less predictable.

Despite the high intensity of light in the tropical and subtropical zones throughout the year, primary production is

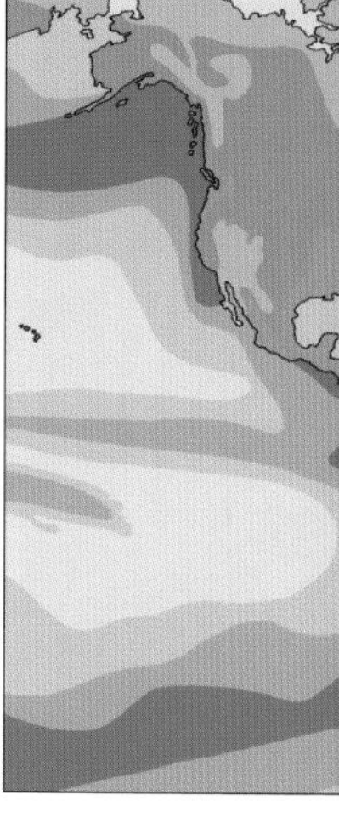

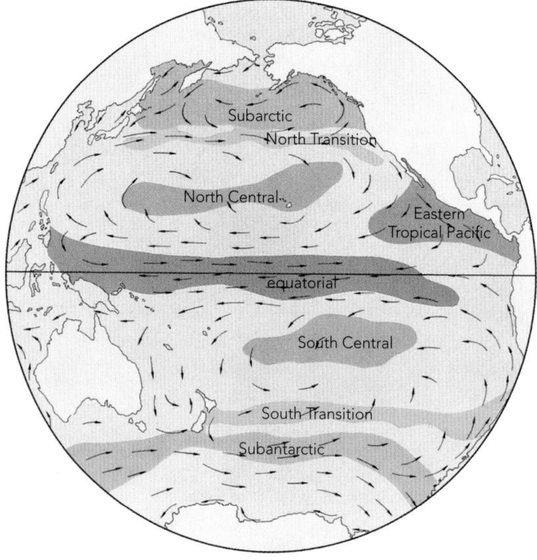

Subarctic assemblage

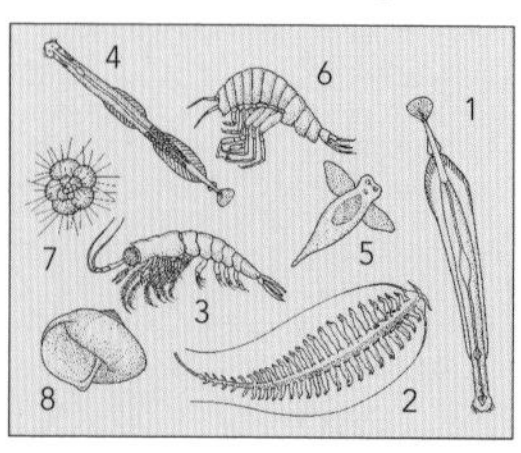

Central assemblage

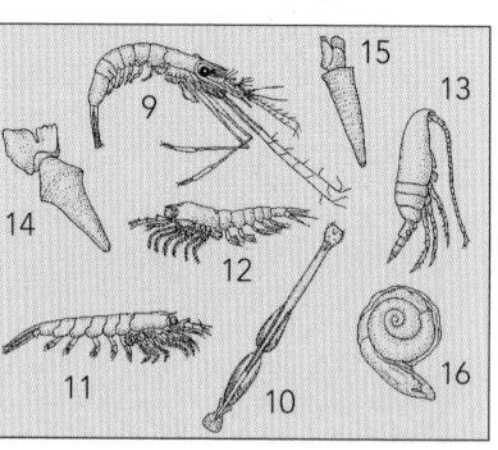

◀ ***The Pacific Ocean*** *has eight major plankton communities, the distribution of which is controlled by the ocean's current systems. The map illustrates the extent of the 'core' zones of the communities and their relationships with the systems: the outermost contour of each zone extends further and may overlap with other zones. The assemblages show zooplankton species common to the subarctic and central Pacific.*

1 *Eukrohnia hamata*
2 *Tomopteris pacifica*
3 *Euphausia pacifica*
4 *Sagitta elegans*
5 *Clione limancina*
6 *Parathemisto pacifica*
7 *Globigerina quinqueloba*
8 *Lamancina helicana*
9 *Stylocheiron suhmi*
10 *Sagitta pseudoserratodenta*
11 *Euphausis brevis*
12 *Euphausis mutica*
13 *Clausocalanus paululus*
14 *Cavolina inflexa*
15 *Styliola subula*
16 *Limancina Lesuerii*

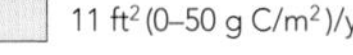

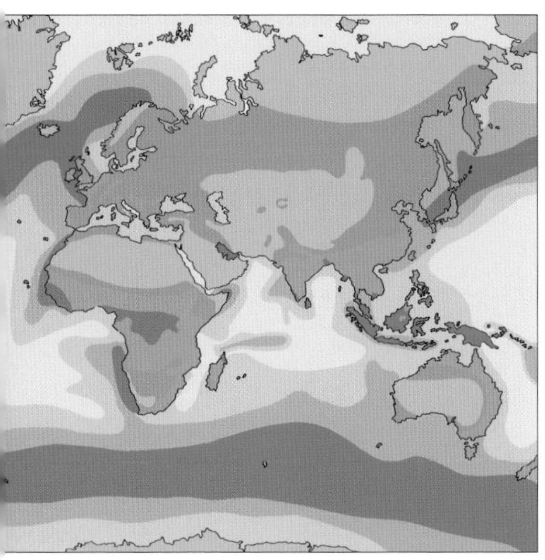

11 ft² (0–50 g C/m²)/yr
11 ft² (50–100 g C/m²)/yr
11 ft² (100–200 g C/m²)/yr
11 ft² (over 200 g C/m²)/yr

11 ft² (0–100 g C/m²)/yr
11 ft² (100–400 g C/m²)/yr
11 ft² (500–800 g C/m²)/yr
11 ft² (over 800 g C/m²)/yr

◀ ***Primary productivity*** *describes the creation of organic matter by plants from inorganic elements, using the Sun's energy captured by photosynthesis. The organic material produced is then available for use by animals.*

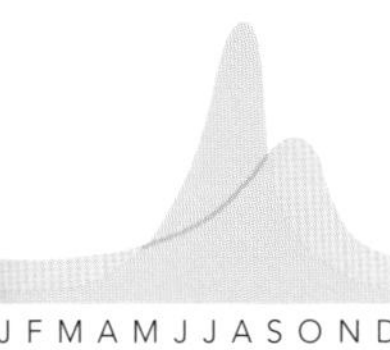

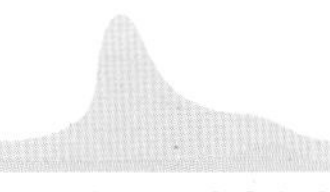

▲ ***Seasonal cycles*** *of plankton production vary according to latitude.*

limited by the concentration of nutrients, especially nitrogen and phosphorus. Tropical and subtropical waters are characterized by low-nutrient concentrations, which limits the growth of phytoplankton. Primary production in the tropics normally is less than a quarter of that of high-latitude regions.

Highly productive areas tend to support larger-sized phytoplankton, and the community of both phytoplankton and zooplankton tends to be composed of fewer species at higher density than in less nutrient-rich areas. In nutrient-deficient areas, individuals tend of be of smaller size, the community is more diverse and individual species occur at lower density. As a consequence, in zones of high-nutrient input, many phytoplankton are large enough to be consumed directly by small filter-feeding, herbivorous fishes. At the same time, the proportion of herbivorous zooplankton tends to be higher in areas of high-nutrient availability. Carnivorous species dominate the zooplankton of nutrient-deficient areas.

Pelagic species and ocean productivity

While the smaller organisms of the surface oceans drift with the surface-water currents, larger, pelagic organisms do not. Fish, such as tuna, for example, undertake lengthy, transPacific and transAtlantic journeys each year, following the seasonally changing patterns of productivity in the surface waters. In addition, species such as the bigeye tuna spawn in areas where the eggs and larval fishes will be carried with the ocean currents and associated plankton community toward areas of high productivity. Whales also migrate over entire ocean basins following the changes in production, and the breeding of many seabirds is timed to correspond with the peak in production of the small fishes on which they prey.

Reproductive strategies in marine animals range from the 'broadcast' spawners, which produce millions of small eggs, to the more conservative species, which produce only a few large eggs. Broadcast spawners rely on the survival of a few individuals from the many millions produced, since most are lost either through predation or through being carried by currents into unsuitable areas. Differences in larval survival from year to year can vary enormously and thus dramatically affect the adult population of these animals.

Species that produce fewer, larger eggs, generally rely on the fact that their young, when hatched, are larger and can avoid the intense predation which occurs in plankton communities. The adults may also protect the eggs, as in the case of anemone fish and many mollusks. These species tend to show smaller fluctuations in numbers, since the mortality rate of their juveniles is comparatively lower.

► ***Marine turtles mate*** *at sea. The male mounts the female and holds on to her carapace with spurs on his flippers. A female can mate with several males and store sperm for successive batches of eggs. Having mated, the females stay in the breeding grounds going on shore at night at about fortnightly intervals to lay several clutches of eggs.*

Mass fertilization

Many sessile, or sedentary, animals are broadcast spawners. In giant clams, each individual is hermaphrodite (having both male and female reproductive cells) and when spawning, the ripe individual first produces sperm, which are released in milky clouds into the water. Chemicals in the sperm then stimulate neighboring clams to begin spawning. Following the production of sperm, eggs are produced and the timing is such that the eggs are released just as neighboring individuals start producing sperm, maximizing the chances of cross-fertilization. A single ripe giant clam may produce many millions of eggs which, if successfully fertilized, will hatch into swimming, planktonic larvae.

Sessile animals use the planktonic larval phase of their life cycle for dispersal. The larvae float passively in the plankton community, where they form the food for larval fish and other

Invertebrate reproduction

Ocean invertebrates reproduce in a variety of ways. Simple animals reproduce by budding, whereby a new individual grows from the body of the original. More complex forms of reproduction involve fertilization of the eggs.

Budding of soft coral

Sea slug laying egg ribbon

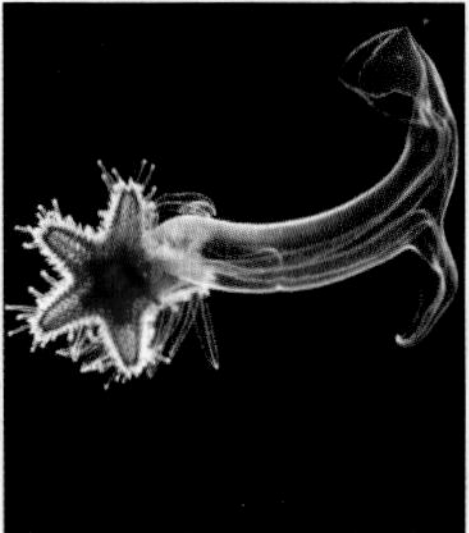

Long-armed sea star larva

predators, such as arrow worms. Since they are carried passively by the currents, they depend on them to bring them into a suitable area for settlement when the larval stage is over. On settling, the larvae of sessile species must choose a suitable site on which to anchor, and site selection is critical to the survival of the adult. If the larvae settle in an unsuitable area, growth may be retarded or the animal may die.

▼ ***Hermaphroditic*** *giant clams reproduce by releasing thousands of sperm and then eggs into the water. The sperm and eggs fuse to develop tiny larvae.*

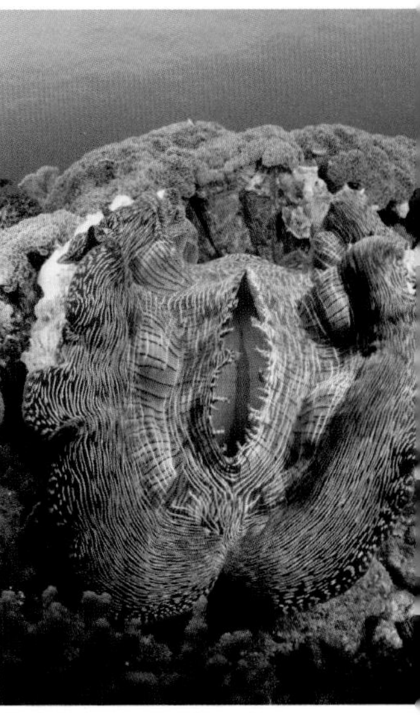

Finding a mate

Reproduction in marine animals involves, first of all, finding a mate, which can present formidable problems, particularly in the deeper ocean where individuals are often widely dispersed. Marine animals live in a three-dimensional world of water, often under conditions of little or no light, and where eyes are of little use. Some animals, therefore, have evolved chemical means of finding and attracting members of the opposite sex. Others, such as the angler fish, have solved this problem in a different way. The males are small, and remain permanently attached to the females.

Several species of coral reef fishes, such as the groupers, form spawning aggregations, returning each year to particular spawning sites where the animals, which are normally widely dispersed over the reef system, produce their eggs and sperm simultaneously. In many island environments, these spawning sites are located in areas where the local current patterns will carry the eggs and larval fishes in a circle, so that by the time they are ready to become reef dwellers, they have been brought back to the vicinity of the reef environment.

INVERTEBRATE MOVEMENT

▲ ***The jellyfish***
Cassiopeia *is commonly found in the warm waters of Florida and the West Indies. In order to swim, it contracts its flattened round bell, which may grow to more than 3 ft (1 meter) in diameter, forcing water out from underneath. It has a jerky upward form of locomotion.*

When compared with vertebrates, the movement of soft-bodied animals is slow, but the diversity of forms of locomotion is considerable, and the structural adaptations of the animals is numerous. Pelagic species are often passive, being carried by the surface water currents or, in the case of animals with floats such as *Physalia*, the Portuguese man-o'-war, drifting with the wind.

Many invertebrate groups in the marine environment are sessile, or sedentary, moving only sufficiently to retreat within a burrow or protective shell, although such movements may be rapid, particularly when a predator threatens the animal. In benthic animals, some form of creeping characterizes many surface living animals, of which the gastropods, or snaillike mollusks, are familiar examples. These animals have a large muscular foot and move by means of waves of contraction which pass along the undersurface of the foot, combined with cilia which operate in a film of mucus secreted by the foot itself. Movement by means of cilia, small hairlike structures, is common in many groups, including the flatworms, which can also swim by means of flapping movements of the body.

The most sophisticated locomotory system is perhaps that of the echinoderms – the starfish, urchins and their relatives. These animals possess small, suckered tube feet, which can be extended forward, attached to the substrate ahead of the animal and then shortened to pull the body along. The brittle stars have developed jointed arms, which can be flexed from side to side in snakelike movements. These arms are covered with short spines, and by curling them around projections on the bottom, the animals can lever themselves over the surface.

Some slowly creeping sea slugs, such as the Spanish dancer, when threatened, open the mantle and move the flaps to swim away. This also displays a vivid pattern of colored eyespots which startles the predator, allowing the animal to make good its escape.

A major advance on the soft-bodied animals is seen in the crustacea, which, like their terrestrial relatives, the insects, have jointed skeletons that can be moved by means

Invertebrate swimming

Comb jellies swim by using comblike plates called ctene (1). Each of the eight ctene is covered with rows of rhythmically beating cilia which propel the animal. The ctene beat away from the mouth (2), thus the animal swims mouth first. Simple muscles provide some control over the tentacles (3).

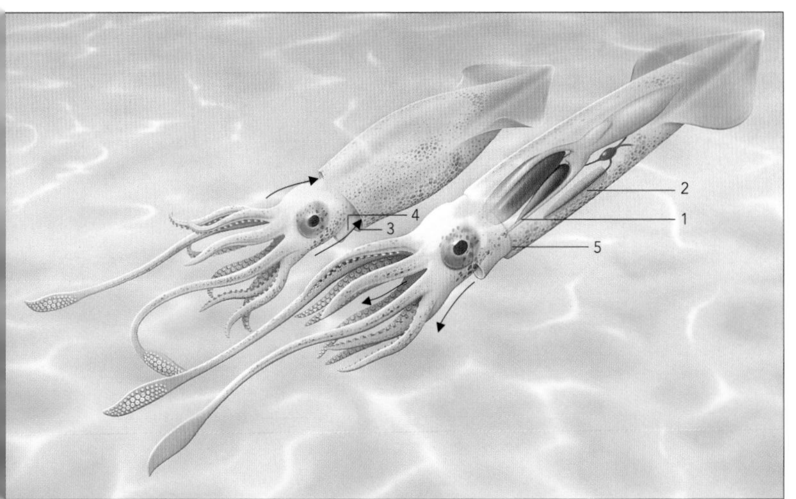

▲ ***The squid**, Pyroteuthis, has a muscular cavity (1) that encloses its gills (2). To propel itself the squid expands the cavity, sucking in water through a wide slit (3). Water cannot flow up the funnel (4) because of a one-way valve. To move, the squid contracts the cavity, forcing water out of the funnel. The overlapping edges of the split (5) ensure the water is forced out in a narrow stream, so creating better thrust.*

of opposing pairs of muscles. These animals have developed a wide range of walking and swimming styles based on the use of jointed legs and swimming paddles. These styles range from the stately walk of the sea spiders to the scuttling of crabs and the frenetic swimming of small copepods, which kick their jointed legs to move in the water column.

Burrowing animals

Locomotion beneath the surface of sediment involves some form of burrowing. Many soft-bodied worms burrow by extension and contraction of the segments of the body. The individual segments of the worm's body contain fluid which is incompressible, so that when the circular muscles of the body wall contract, the longitudinal muscles relax and the segment is extended, since the overall volume cannot change. This forces the forward end of the body into the sediment. Then contraction of the longitudinal muscles causes relaxation of the circular muscle, and the segment becomes short and fat, anchoring the animal in its burrow. Sequential contractions passing along the length of the animal's body move it through the sediment. Locomotion in this way requires a lot of energy, and because of this, numerous burrowing animals live and rest in permanently constructed burrows.

▶ ***Mollusks** such as this whelk, Baccinum sp., move by means of waves of muscle contraction which pass along the foot. Locomotion is aided by hairlike cilia and mucus, secreted by glands on the undersurface of the foot.*

The main mode of locomotion among marine vertebrates is swimming, especially powered swimming, which depends upon a source of forward thrust, the ability to steer and to change course, and the ability to stay afloat. Aquatic vertebrates have achieved large size, in part due to the buoyancy or support provided by the water. Nevertheless, animals are denser than water, and thus have a tendency to sink unless they counteract this by swimming upward. Sharks and rays reduce their density by storing oils in the liver. They will still sink unless they swim continuously.

A major advance in the teleost, or bony fishes, is the development of the swim bladder – an air- or gas-filled sac lying below the vertebral column, which results in neutral buoyancy. Gas can be secreted into, or reabsorbed out of, the gas bladder according to the depth of the fish, and a teleost fish with neutral buoyancy neither sinks nor rises to the surface. This allows the animal to conserve energy, as it maintains its position merely by small adjustments of the fins that counteract the movements of the surrounding water.

Types of fins

Since water is constantly moving and animals need to change direction, there is a need for stabilizing structures and steering devices. The median fins of fish, which lie in the dorsal and ventral midline of the body, provide a broad surface that prevents rolling, and these fins can be extended or retracted as required. The paired fins, the pectoral fins at the sides of the body and the pelvic fins which lie further back and more ventral, are those used in steering and in four-fin braking. They also prevent pitching, which would result in an up and down movement of the front of the body. By holding the paired fins extended and at an angle to the water the fish maintains its stability. Some teleost fish extend all four paired fins simultaneously, presenting a large surface area to

▼ ***True bony fishes*** *are diverse in shape and type of locomotion. They all possess a tail with equal upper and lower lobes providing solely horizontal thrust. Bony fish achieve neutral boyancy by adjusting their density using the swim bladder. The fish can expand or contract the bladder by secreting gas into or absorbing gas out of it, so adjusting the volume and external pressure and counteracting the tendency to sink or float to the surface.*

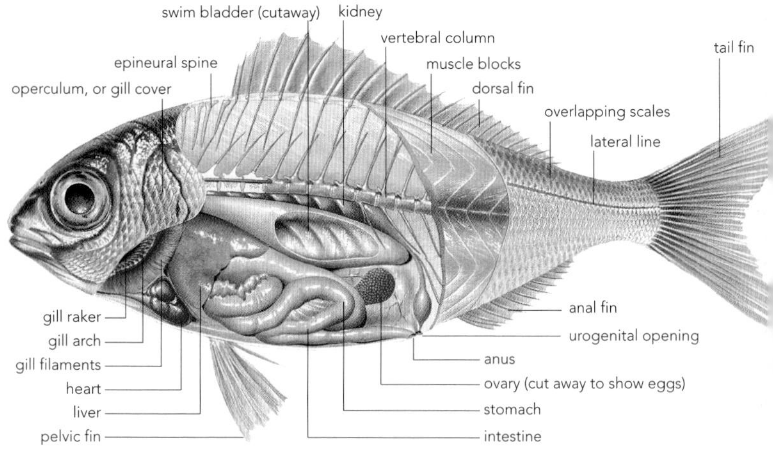

▲ ***Brightly colored fish*** *swim around a coral reef in the Gulf of Aqaba, the Red Sea. The smaller fish that inhabit reefs do not require the same strength and speed as some of the larger fish and swim with their fins rather than the tail.*

▲ ***The hatchet fish**,* Argyropelecus, *inhabits the deep sea where food is scarce. Its mouth can gape widely, enabling it to seize large prey.*

▲ ***The tuna*** *is a muscular, long-distance swimmer whose capacity for speed is indicated by its torpedo shape, lunate tail and pectoral fins.*

▲ ***The flattened body*** *and camouflaged pattern of the plaice are adaptations for life on the seabed. Larval plaice are shaped like normal fish.*

▲ ***Marlins*** *are aggressive, fast-swimming predators which can achieve a length of more than 14 ft (4 m). The spearlike bill is used to maim prey as it swims through shoals.*

the oncoming water and thus braking over short distances. Other bony fishes use the paired fins like paddles for locomotion and delicate movements in confined spaces, such as the crevices on a coral reef.

Locomotion using fins rather than tails as the source of forward power, is found in a variety of bony fishes with rounded body shapes and in the rays which have large pectoral fins or wings which are moved up and down in the water providing both lift and forward thrust. Penguins swim underwater using modified wings that serve as paddles, while forward thrust in turtles is achieved by the paddlelike forelimbs that are moved forward and upward before being drawn backward and downward through the water. The movements of the forelimbs are very similar to the movements of a bird's wing when flying. The airfoil shape of the forelimbs reduces the resistance as they are moved forward, and by rotating the limb on the down stroke, a large surface area is presented to the water increasing the forward thrust.

Water as a medium presents resistance to movement, and the efficiency of swimming depends upon reducing turbulence in the water as the animal moves forward. A smooth outer surface is therefore a prerequisite for high-speed swimming, and marine mammals have lost their hair while the scales of fishes present a smooth surface over which the water can flow uninterrupted. Swimming using paired fins or paddles is generally slower than using the tail, and to achieve greater speeds and more efficient movement, fast-swimming fishes such as the tuna have highly streamlined, torpedo-shaped bodies that result in a laminar flow of water

▲ ***The seahorse*** (Hippocampus) *lacks a caudal (tail fin) and relies instead upon its dorsal (back) fin to move forward through the water. The fin beats 20–30 times per second. The pectoral fin controls its turning direction. Seahorses regulate whether then swim up or down by controlling the volume of gas in their bodies.*

over the body surface. The paired fins are held close to the body so that no resistance to forward movement is presented, and the animal slips through the water at high speed. Continuous high-speed swimming demands a lot of energy and hence a plentiful oxygen supply to the muscles as they contract. Tunas have developed red muscle along the sides of the body, containing a large reserve of oxygen for use by the muscles during long-distance swimming.

Very few invertebrates are active swimmers, although some crustaceans swim using modified paddlelike limbs. The most efficient invertebrate swimmers are the squid and cuttlefish. The cuttlefish has a continuous fin fold around the flattened disc-shaped body along which pass waves of movement providing forward thrust. In contrast, the fins of squid are reduced to triangular stabilizers at the rear of a torpedo-shaped body. The squid moves backward through the water by contraction of the entire body, which forces a jet of water out through the siphon, a series of quick spurts rather than a continuous forward motion.

Types of tails

To achieve efficient forward movement, an actively swimming animal requires a source of power located behind the center of mass of the animal, hence most active swimmers have developed some form of tail. Having the source of power behind the center of mass gives greater control over the forward movement than dragging the body through the water. Tails vary in shape from the horizontal flukes of the whale to the homocercal (symmetrical) tail of the teleost fish, which has lobes of equal size above and below the midline of the body. In contrast, the tail of sharks is heterocercal – the dorsal lobe is larger in size than the ventral lobe. Such a tail not only provides the power for forward movement, but also provides lift to the rear of the animal. This lift counteracts the tendency of the animal to sink but also causes the tail to rise relative to the head. The paired pectoral fins at the front of the shark are held at an angle to the water, providing lift to the front, which balances the lift at the rear provided by the tail.

Since the teleost fish achieve neutral buoyancy by means of the swim bladder, the need for upward movement is reduced, and both lobes of the tail are equal in area. The fossil reptilelike Ichthyosaurs had tails which were the reverse shape to those of sharks; the ventral lobe was greater in area than the dorsal one. In these animals, which were air breathing, the problem was one of floating to the surface. Their tail shape was an adaptation to provide a downward thrust, enabling the animal to move more easily below the water surface.

Evolution of swimming

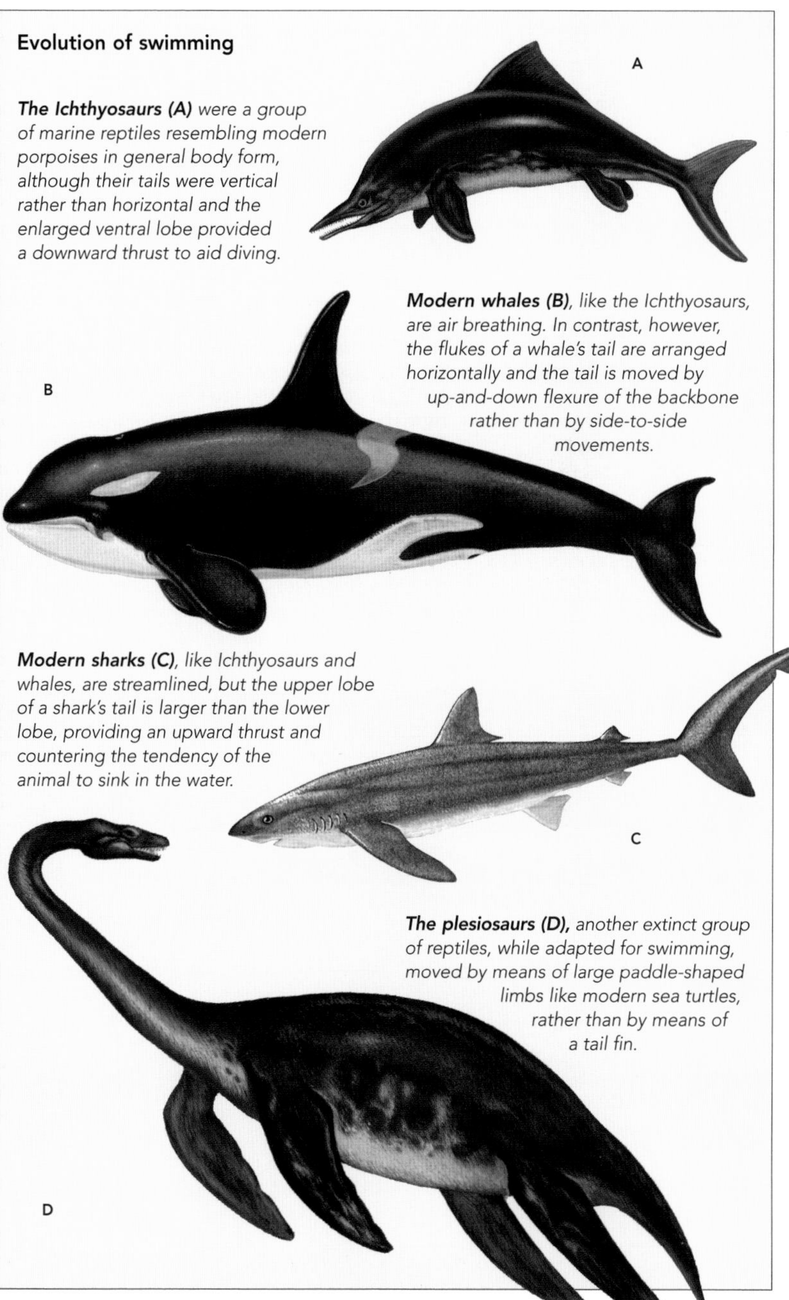

The Ichthyosaurs (A) *were a group of marine reptiles resembling modern porpoises in general body form, although their tails were vertical rather than horizontal and the enlarged ventral lobe provided a downward thrust to aid diving.*

Modern whales (B)*, like the Ichthyosaurs, are air breathing. In contrast, however, the flukes of a whale's tail are arranged horizontally and the tail is moved by up-and-down flexure of the backbone rather than by side-to-side movements.*

Modern sharks (C)*, like Ichthyosaurs and whales, are streamlined, but the upper lobe of a shark's tail is larger than the lower lobe, providing an upward thrust and countering the tendency of the animal to sink in the water.*

The plesiosaurs (D), *another extinct group of reptiles, while adapted for swimming, moved by means of large paddle-shaped limbs like modern sea turtles, rather than by means of a tail fin.*

▲ ***Feather stars*** *are bottom-dwelling animals that feed on detritus which sinks from the surface, or capture their food from the water.*

In its simplest form, a food chain can be seen as a series of links with each species in the chain depending on the species below as a source of food energy. At the bottom of the chain lie the primary producers, or autotrophs, and the growth of individuals and the increase in their numbers provides the food source for herbivores, which are in turn eaten by carnivores in the level above.

All life processes involve the expenditure of energy, so not all the energy produced by autotrophs is available for consumption by the herbivores. Some energy is lost through heat, and some phytoplankton die before they are consumed. Actively moving herbivores and carnivores use energy merely in locomotion, and animals do not totally digest all the food they take in; some energy is lost through fecal material. Dead animals and plants, and fecal matter provide the energy for decomposing bacteria and for sedentary, filter-feeding animals on the ocean floor that feed upon the rain of organic material dropping down from above.

Marine food webs

On land, simple pyramids of numbers, or biomass (the weight of living tissue), demonstrate the decline in energy as one passes up a food chain. The numbers, or biomass, of primary producers exceeds that of the herbivores, which in turn is greater than that of the carnivores. Marine communities are, however, quite different: they display inverted pyramids in which the numbers and biomass of primary producers are less than those of the herbivores. If, however, the production over an entire annual cycle is considered, then the production of the autotrophs exceeds that of the herbivores. This reflects the rapid rates of production of individual phytoplankton, which reproduce in a few hours or days, compared with the

1 Phytoplankton
2 Mollusk larva
3 Decapod larva
4 Copepod, *Acartia* sp.
5 Barnacle larva, *Balanus* sp.
6 Chaetognatha, *Sagitta* sp.
7 Pterapod, *Limacina* sp.
8 Sand eel larva, *Ammodytes* sp.
9 Copepod, *Acartia* sp.
10 Copepod, *Temora* sp.
11 Copepod, *Calcanus* sp.
12 Euphausiid, *Nyctiphanes* sp.
13 Cladocera, *Evadne* sp.
14 Cladocera, *Podon* sp.
15 Hyperiid amphipod
16 Appendiculate, *Oikopleura* sp.
17 Herring, 0.25–0.5 in (0.5–1 cm)
18 Herring, 1.25 in (3 cm)
19 Herring, 1.5 in (4 cm)
20 Herring, 1.5–5 in (4–12 cm)
21 Adult herring, to 16 in(40 cm)

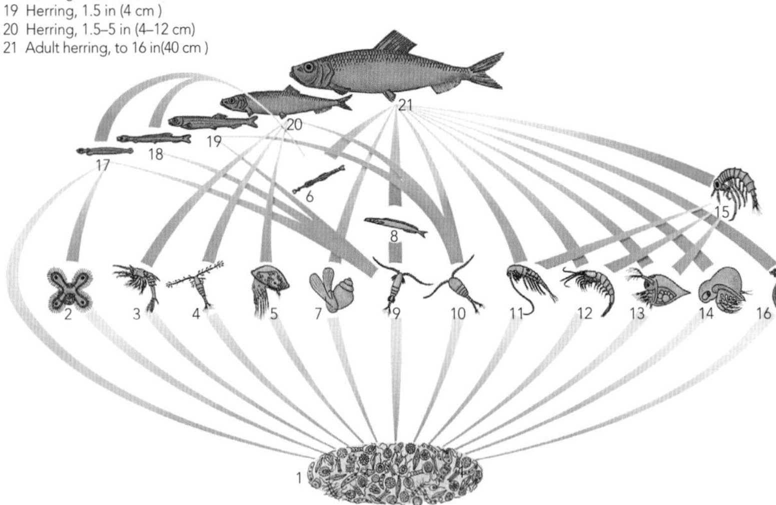

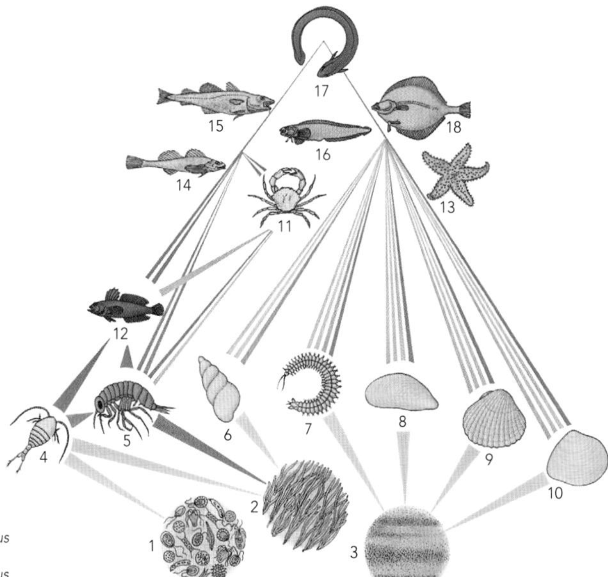

▶ ***This complex food web**, based on the community of Danish coastal waters illustrates the kinds of interrelationships which exist in shallow-water communities.*

1 Planktonic algae
2 Large plants (*Zostera*)
3 Detritus
4 Copepod
5 Small crustacea
6 Gastropod, *Hydrobia* sp.
7 Worm, *Neresis* sp.
8 Mussel, *Mytilus* sp.
9 Cockle, *Cardium* sp.
10 Tellin, *Macoma* sp.
11 Shore crab, *Carcinus* sp.
12 Goby, *Gobius* sp.
13 Starfish, *Asterias* sp.
14 Sea scorpion, *Taurulus bubalis*
15 Cod, *Cadus morhua*
16 Eelpout, *Zoardces vivparus*
17 Eel, *Anguilla anguilla*
18 Flounder, *Platichthys flesus*

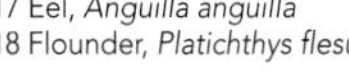

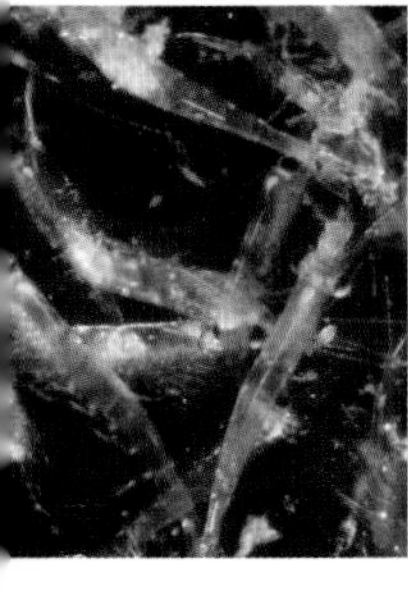

▲ ***The Antarctic krill** is a carnivorous member of the plankton community. Baleen whales eat up to two tons of krill a day.*

◀ ***The feeding habits** of the herring change during the different stages in its life. As the animal grows, it is capable of feeding on larger-sized prey and the range of different species taken increases.*

zooplankton, which may take weeks or months to reproduce, or the smaller predatory fishes, which take a year or more to produce new individuals.

Complex marine communities rarely consist of a simple chain of single species, with each feeding on the species in the level below. Such communities would be extremely unstable and liable to fluctuate considerably in numbers, if not collapse entirely, if a single link in the chain should disappear. In practice, therefore, food webs, rather than food chains, exist, with a complex interlinkage between different members of the same community. A single species such as the flounder, for example, may feed on several different species of annelid worms and mollusks. It may compete for some of these resources with predatory starfish, eels and other fishes. To avoid competition, different species have devised different strategies to divide the available resources between them. The filter-feeding bivalves, for example, which are dependent on the suspended detrital material in the water, will select different particle sizes or will position themselves in different locations to make use of different sections of this resource.

Overfishing represents one of the greatest threats to the stability of marine ecozystems, since many fishing techniques not only damage and destroy the environment, but the links in the complex web of feeding relationships are disrupted by the selective removal of certain species. Regrettably, our understanding of marine food chains is not great enough to enable us to predict the impact of overfishing on other species in marine communities.

A major feature of the marine environment is the small size of primary producers, and many marine organisms have consequently developed a mode of feeding that is unique to the aquatic environment: filter-feeding.

Filter feeders

Filter feeders ingest suspended particles, which are generally much smaller than the animals that feed on them. One of the most extreme contrasts in size is that between the blue whale, which can reach lengths of more than 100 feet (30 meters), and its food, the krill, a crustacean, which is only 1 inch (2.5 centimeters) long. The whale filters these small animals from the water as it swims, a mechanism that is used by many planktivorous fish. On a smaller scale, herbivorous copepods filter-feed by means of the rowing action of their limbs, which are fringed with hairs that strain the minute phytoplankton from the water. The copepods are themselves fed on by Euphausiid shrimps, which comb them from the water in the same way.

Sedentary animals may use a more passive mode of food collection. Feather stars, for example, produce mucus strands or threads which float and trap suspended particles, before they are drawn in and rolled down to the mouth by tube feet which line the arms. Animals such as sponges, salps, many bivalve mollusks and sea squirts, draw water into their body by means of cilia, which beat inward, drawing water into an internal chamber where food is separated from the water current.

Detritus feeders

Fecal matter and dead marine organisms provide the energy source for a complex community of benthic organisms, of which the bacteria are an important link. The surface sediments of the ocean floor contain large numbers of bacteria, which serve as a food source for small interstitial animals adapted to moving between the sand grains. Effectively, the sediment represents a mixed organic and inorganic soup

Ways of feeding

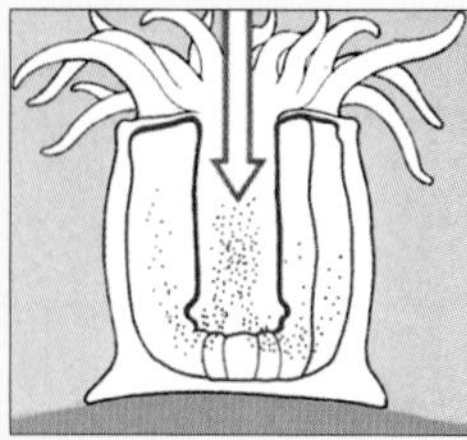

▲ **Condylactis passiflora**, *a sea anemone, catches its prey with the stinging tentacles that surround the mouth.*

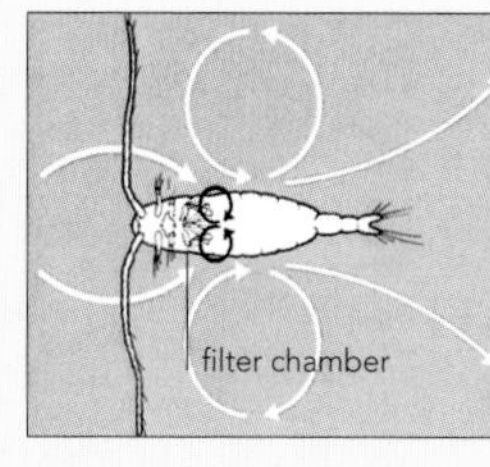

▲ **Calamus finmarchicus**, *a copepod, extracts tiny food particles from the water by means of a fringe of hair on its limbs.*

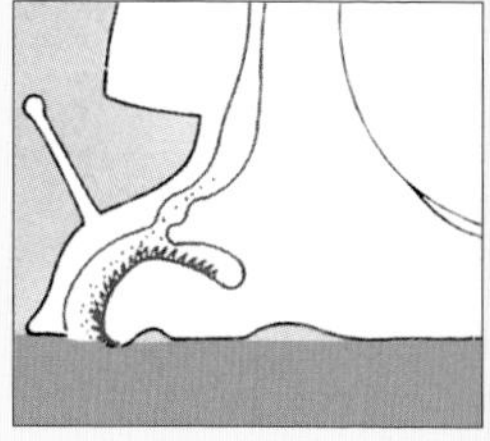

▲ ***Herbivorous sea snails*** *scrape their algal food from the surface using a tough, filelike radula moved by a muscular tongue.*

◀ ***The large size*** *of baleen whales in comparison with their prey, krill, means that they must filter their prey from the water.*

which is consumed directly by many burrowing organisms that digest the organic materials and pass out a stream of clean inorganic particles.

Many crabs, including the fiddler crabs, pick up individual sediment particles and brush them clean with their setae, hairlike structures on the mouthparts, retaining the edible materials and discarding the sediment. Some detrital-feeders, such as brittle stars, sweep the area in front of their retreat with their arms. Modified tube feet on the undersurface of the arms produce a sticky mucus, which is used to collect food particles.

Grazers

Grazing is restricted to shallow coastal waters, where single-celled algae grow over the surface of the bottom, and larger seaweeds and seagrasses provide a source of larger plant food. Many species have developed mechanisms for scraping this plant production off the surface of the rocks. The radula of mollusks consists of small teeth arranged in rows over the surface of a muscular tongue that is rasped over hard surfaces, removing the algae. Sea urchins have five teeth that can be rotated in and out and used, like the radula, to scrape unicellular algae from rock surfaces. A variety of small herbivorous fish crop the algae that forms a short turf in many tropical reef systems. They are important in maintaining diversity by controlling the growth of the algae, preventing them from smothering the delicate, juvenile corals.

▼ ***Surgeonfish*** *are grazers that feed on the algae that covers coral reefs.*

A wide range of sessile (sedentary) animals live in shallow waters, providing the opportunity for predation, which resembles more the grazing of terrestrial herbivores than the active chase and capture associated with predator-prey interactions on land.

Corals are 'grazed' by crown-of-thorns starfish which evert their stomachs over the surface of the coral colony, digesting out the polyps from the hard, calcareous skeleton. In contrast, the strong jaws of the parrot fish crunch off coral branches, which are then ground up by modified platelike teeth at the back of the jaws. Many smaller animals browse on coral colonies by delicately extracting individual polyps from inside their protective skeletons. Some sea slugs even store the stinging cells collected from their coral prey and use them for protection against other predators.

Strong, calcareous skeletons, into which the soft-bodied animals can retreat, are no guarantee of protection. Bivalve mollusks, for example, are still vulnerable to predation since many marine snails have modified their radula to form a drill, which is used to pierce a neat circular hole through the shell. Once through the shell, the gastropod secretes digestive enzymes into the mussel, clam or oyster, and sucks out the resulting protein soup.

Active hunters

Fish predators display classic hunting strategies, and some, such as thresher sharks, will hunt in packs, using the long whiplike tail to herd schools of mackerel and herring into tight shoals, before moving quickly in to feed. Active predators often hunt by sight, and the coloration of potential prey is designed to reduce their visibility. In the lighter surface waters, many fish species are counter-shaded, with darker

▼ ***Many surface schooling** fish, such as these mackerel, are counter shaded with darker dorsal surfaces and lighter, silvery undersides. This pattern of counter shading provides protective camouflage from above and below. The silver lower surface blends in with the sunlight from above, making it harder for predatory fish swimming below to spot them, while the darker dorsal surface blends with the deep blue of the ocean water providing protection from seabirds.*

▲ ***The great white shark*** (Carcharodon carcharias) *is among the most fierce of active predators. Sharks have an acute sense of smell and are attracted to wounded or dying fish. As sharks cannot chew, they tear chunks from larger prey by grasping a mouthful and then twisting in the water to gouge out a piece of flesh. Their razor sharp teeth are set in rows parallel to the jaw and are replaced from behind as the front ones are lost or damaged.*

blue or gray dorsal surfaces and pale or silvery bellies. This reduces their visibility when viewed from below against the sunlit surface and when viewed by seabirds from above against the darker ocean.

Squid, cuttlefish and octopus are all predatory animals, and while octopus often sit and lurk in dens waiting for unsuspecting snails and fish to approach, squid and cuttlefish are active predators of crustacea. Some species of squid will direct a gentle flow of water onto the sand surface to expose hidden shrimps, which are then seized with an elongated pair of tentacles. Cuttlefish, squid and octopus, when threatened by larger predators, will avoid capture by expelling a large cloud of ink as a distraction, darting rapidly away and immediately changing their color pattern to confuse the predator.

Patient predators

Sit-and-wait predators are also numerous, particularly at depth, where the numbers of potential prey are few. Such predators are often camouflaged, relying on the inability of the prey to detect their presence to bring them within reach of capture. The deep-sea angler fish encourages smaller fish to swim within striking distance by means of a lure, a small structure on the angler's head which is either luminescent or may resemble a small worm. Prey approaching in search of a meal find themselves under attack.

As a defense against predation, many species of marine organisms have developed toxic or noxious chemicals and adopted brightly colored patterns to warn potential predators of their characteristics. Many of the venomous sea snakes are striped black or dark blue and white, although this does not prevent them from being a major item in the diet of tiger sharks.

Since all life evolved in the sea, the invasion of land led to the evolution of numerous adaptations: for supporting the body weight in air; for locomotion using limbs, not fins; for breathing air; for producing eggs or young resistant to desiccation; for conserving body moisture; and for coping with the wide range of diurnal and seasonal temperature extremes experienced on land. Having successfully adapted to such conditions, it is hardly surprising that a limited number of mammal groups have made the transition from the land back into the marine environment.

Some marine mammals, which have made the transition back to the sea only relatively recently, show few major structural differences from their land-based relatives. The marine otters, for example, spend their lives in kelp beds, feeding on mollusks, which they open by smashing against an anvil stone held on their chest. Even though these animals sleep in their kelp-bed habitats, they show little difference from their amphibious cousins, the freshwater otters.

The polar bear, like the otters, has retained its mammalian characteristic of hair. However, hair is an inefficient insulator in water, since it becomes wet and water penetrates to the surface of the skin. True marine mammals, such as the seals, whales and dolphins, rely on the subcutaneous fat or blubber for insulation.

Although the fur seals have retained their fur, they also have a thick layer of fat for insulation, but most seals and sea cows have retained only a sparse covering of body hair and sensory bristles around the mouth. Seals are of two groups, the eared and earless seals, while the Arctic walrus is considered to be separate from both groups. In the eared seals, the hind limbs are held together when swimming and moved in an up-and-down manner like the flukes of whales.

▼ ***Elephant seals*** *derive most of the insulation needed to maintain their body temperature when diving from the thick layer of blubber beneath the skin.*

Underwater expertise

Marine mammals dive and swim easily, but their capacity for extended dives varies considerably.

▲ ***The marine otter*** *remains at sea for most of its life and can stay submerged for minutes.*

▲ ***The California sea lion*** *has learned to dive to depths of 820 ft(250 m) in the Pacific.*

▲ ***Seals*** *make short dives lasting about 10 minutes, but the Weddell seal can dive to 2000 ft (600 m).*

▲ ***The walrus*** *dives to 330 ft (100 m) remaining submerged for 10 minutes to collect mollusks.*

▲ ***The manatee*** *dives in shallow water to graze on seagrasses and returns to the surface constantly.*

▲ ***The Arctic polar bear*** *swims using its powerful limbs but spends much of its time on land or on the ice. While the long fur is an excellent insulator on land, it provides little protection against the cold when wet. It is likely that the polar bear adopted its semimarine existence relatively recently.*

On land, the limbs can be rotated forward, and the animals move by means of a clumsy but nevertheless effective gallop. In contrast, the hind limbs of the earless or true seals cannot be rotated forward, and on land these animals must drag themselves along using the short forelimbs.

Seals come on land to breed and have their young which, in the case of earless seals, are born in such an advanced stage of development that in some species they can swim within a few hours of birth. In contrast, the young eared seals are less well developed at birth and must spend a long time on land before taking to the water. These differences between the two groups suggest that the earless, or true, seals returned to the sea before the eared seals.

The sirenians, or sea cows, which include the dugong and the manatee, inhabit estuarine and shallow coastal seas in the tropics and are highly adapted for an aquatic existence. The skin is generally hairless, the hind limbs have been lost and the end of the body has developed horizontal tail flukes like those of whales and dolphins. Sea cows never come onshore. They are vegetarians, grazing on seagrass meadows, and live in social groups or herds. Unlike the whales and dolphins, however, they are unable to dive for extended periods and must return to the surface on average every five minutes to breathe.

Although reptiles dominated the marine environment during the Mesozoic era, they are now restricted to seven species of marine turtle, several hundred species of sea snakes, one species of marine iguana, and a few estuarine crocodilian species.

The turtles have horny beaks and modified front limbs, which serve as the main source of power for locomotion, the rear flippers being used only for steering. The forelimb is paddle-shaped, and powerful thrusts of these limbs provide the forward movement – a form of swimming also found in the extinct plesiosaurs, which became extinct at the same time as the dinosaurs on land. The streamlined shape of the turtle's carapace provides a smooth surface over which water flows in layers, preventing turbulent currents, which would interfere with efficient swimming. In contrast to the land-based tortoises, sea turtles have reduced the weight of the skeleton and are much more flattened. They come ashore to breed, laying eggs in excavated nests at the head of beaches. Many of the young are lost to predators on their journey from the nest to the sea, and a more recent problem has been the inland movement of hatchlings attracted to the lights of tourist hotels, which are a stronger stimulus than the moonlight ocean.

The marine iguana, *Amblyrhynchus*, of the Galapagos Islands is really an amphibious lizard spending a considerable portion of its time on land. In general body form, these animals are very similar to their terrestrial relatives, although they swim and dive, and feed on seaweeds. Like seabirds, they have salt-excreting glands in the nostrils, in contrast to turtles, which excrete excess salt through ducts in the eye socket.

▼ ***The marine iguana,*** Amblyrhynchus cristatus, *is the only truly marine lizard. Because the waters it inhabits are cold, it basks in the sun to raise its temperature.*

► ***The sea snake***, Laticauda colubrina, *has a flattened paddle-shaped tail to aid in swimming. Unlike most sea snakes, which never come on land and give birth to young at sea, this species prefers to stay close to shore.*

The saltwater crocodile is estuarine rather than truly marine, although subadults will move long distances along the coast in search of a suitable territory. An early missionary account from Fiji records the killing of a saltwater crocodile that must have crossed from the Solomon Islands. Once common throughout southeast Asia, Melanesia and Australia, these animals have been greatly reduced in numbers and in range by being hunted for their skins.

The sea snakes are more adapted to the marine environment than the iguana and crocodile, since all species have a flattened, paddle-shaped tail. Unlike terrestrial snakes, the belly scales of true sea snakes are much smaller, similar in size to the dorsal scales, since they do not crawl. Sea snakes are a diverse group of highly venomous, marine animals, the vast majority of which occur in the shallow waters of the southwest Pacific Ocean, between Malaysia and Australia. Sea snakes are related to the cobras and kraits on land. The tail is laterally compressed, forming a paddle, and the animals swim like eels by lateral flexure of the body. *Pelamis* is completely pelagic and feeds on small fish which it catches by means of a lateral strike. Many of the smaller-headed species of sea snake feed on fish eggs, while the genus *Laticauda*, the sea kraits, feed on moray eels on coral reefs. Most sea snakes are ovoviviparous, the eggs hatching inside the female's reproductive tract and the three to eight young emerging as fully formed, juvenile snakes. Birth takes place at sea and, with the exception of the *Laticaudine* sea snakes, no species comes on land.

Among the marine mammals, whales and dolphins are the most specialized, with physical adaptations to the marine environment that include the loss of hair and hind limbs, and the development of powerful horizontal tail flukes moved up and down by flexure of the vertebral column. These specializations are now so complete that whales and dolphins are incapable of surviving out of water. Their degree of adaptation suggests that the group must have returned to the sea around 65 million years ago, at the beginning of the Tertiary period.

The diving ability of whales is remarkable considering that they are air-breathing. Before a deep dive, a whale empties its lungs completely, thus eliminating any risk of the bends. To reduce oxygen consumption during the dive, only the heart and brain receive a continuous supply of oxygenated blood. The powerful swimming muscles of these animals are rich in oxygen, stored in myoglobin – a red pigment similar to the hemoglobin of blood – that provides oxygen to the muscle tissue during the dive.

Living whales and dolphins can be divided into the *Odontoceti*, or toothed whales, and the *Mysticeti*, or baleen whales,

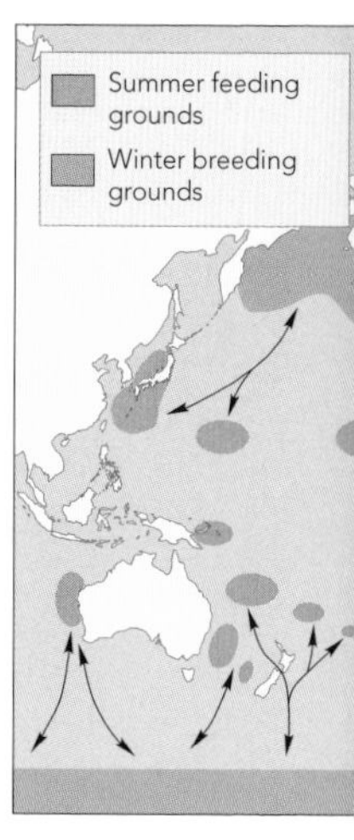

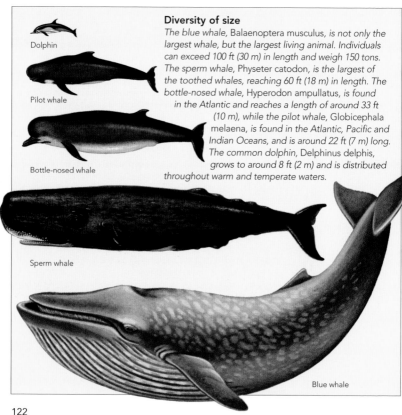

Diversity of size

The blue whale, Balaenoptera musculus, *is not only the largest whale, but the largest living animal. Individuals can exceed 100 ft (30 m) in length and weigh 150 tons. The sperm whale,* Physeter catodon, *is the largest of the toothed whales, reaching 60 ft (18 m) in length. The bottle-nosed whale,* Hyperodon ampullatus, *is found in the Atlantic and reaches a length of around 33 ft (10 m), while the pilot whale,* Globicephala melaena, *is found in the Atlantic, Pacific and Indian Oceans, and is around 22 ft (7 m) long. The common dolphin,* Delphinus delphis, *grows to around 8 ft (2 m) and is distributed throughout warm and temperate waters.*

Dolphin

Pilot whale

Bottle-nosed whale

Sperm whale

Blue whale

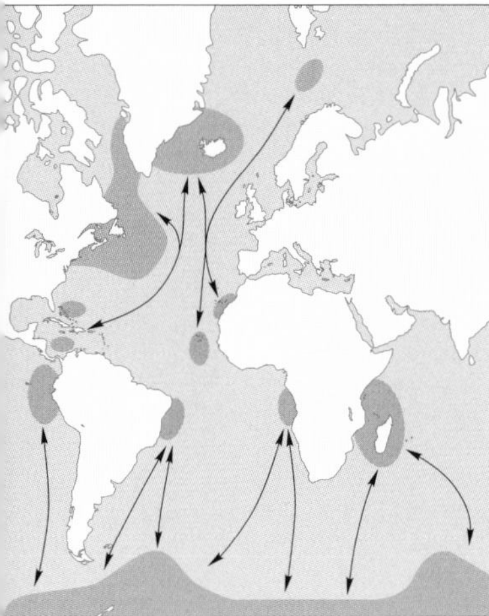

▲ ***Migration of humpback whales*** *is generally north-south throughout all the world's oceans. During the summer months, the humpbacks remain in high-latitude, plankton-rich feeding grounds, heading to their warmer-water breeding areas during the winter months.*

which have lost the teeth and replaced them with horny, fringed plates suspended from the upper jaw. When feeding, the baleen whales swim through the plankton with the mouth open, then close the jaws and raise the tongue and floor of the mouth, forcing water out through the sides of the jaws and trapping the krill inside the mouth. A single mouthful may contain no more than a few pounds of food, but a large whale feeding in this way may harvest 2 tons of krill a day.

In contrast, the toothed whales often have large numbers of rather simple, conical teeth and they are active predators of fish and squid. Some have reduced the numbers of teeth to a single pair, while others may have as many as 200. The largest of the toothed whales is the sperm whale, which can reach 60 feet (20 meters) in length and which dives to considerable depths to capture the giant squid on which it feeds. Smaller whales and dolphins may use a sonar system to detect their prey and for navigation in low light intensities. High-frequency sound, in the form of clicks, is beamed out through the melon, a raised lumplike structure of the forehead, and reflected sound is detected through the enlarged lower jaw.

The complex songs of humpback whales are believed to be part of a signaling system, which enables individuals to detect each other at large distances and to recognize members of the same family unit. Whales and dolphins are highly social animals occurring in family units, with members of the group sharing responsibility for protecting the young, which are born singly, tail first. Gestation is long, around 11 months for baleen whales and more than a year for toothed whales. The young are born at an advanced stage of development, since they must swim immediately to the surface to take their first breath. Early growth is rapid; the fin whale calf, for example, which is around 20 feet (7 meters) long at birth, measures around 40 feet (14 meters) by the time it is six months old.

► ***The humpback whale****,* Megaptera novaeangliae, *is about 50 ft (15 m) long and less streamlined than other whales. It is remarkable for its long narrow flippers which measure up to one-third of the entire body length. These animals harbor a variety of encrusting animals such as barnacles.*

MARINE BIRDS

Probably the most truly oceanic of the 285 species of seabirds are the albatrosses, some species of which may spend up to nine months at sea and which cannot walk on land. Their narrow wings are of considerable length in proportion to their width and are too inflexible to maintain flapping flight, making these birds extremely efficient gliders. By turning upwind, they obtain maximum lift, rising effortlessly into the air or turning downwind to achieve a powerful dive. The stronger the wind, the more efficient their flight, and during conditions of calm weather, they may have difficulty in taking off. The petrels, a group of widespread smaller species, have a different form of flight but are no less accomplished, skimming above the surface of wave crests or turning into the comparative calm of troughs to snatch food from the surface.

In addition to the true seabirds, numerous ducks, geese, divers and wading birds inhabit the coastal areas of the world, feeding extensively in estuarine areas where long-

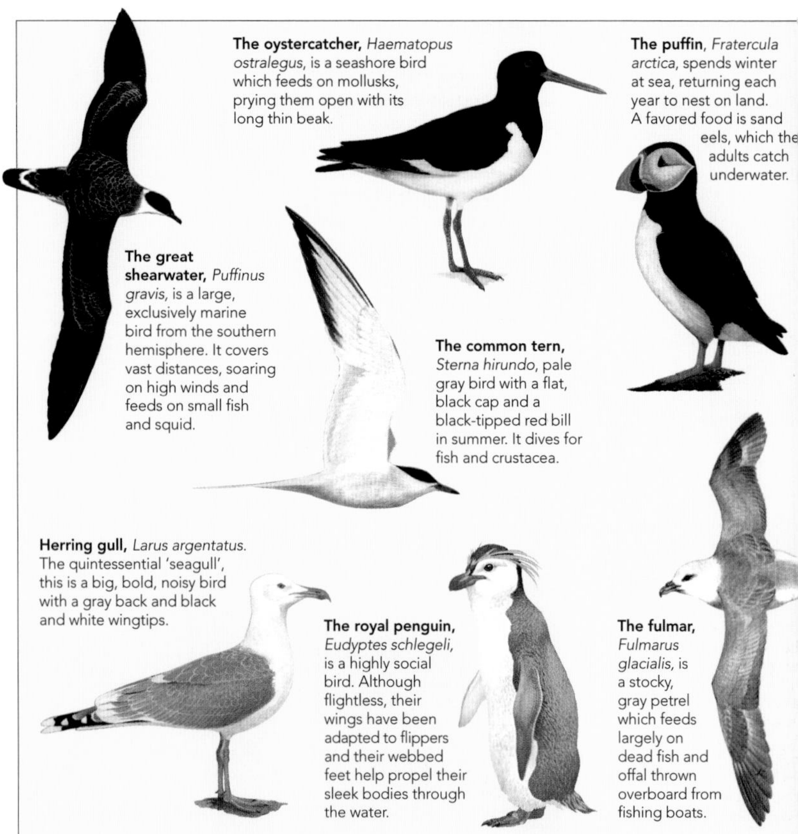

The oystercatcher, *Haematopus ostralegus*, is a seashore bird which feeds on mollusks, prying them open with its long thin beak.

The puffin, *Fratercula arctica*, spends winter at sea, returning each year to nest on land. A favored food is sand eels, which the adults catch underwater.

The great shearwater, *Puffinus gravis*, is a large, exclusively marine bird from the southern hemisphere. It covers vast distances, soaring on high winds and feeds on small fish and squid.

The common tern, *Sterna hirundo*, pale gray bird with a flat, black cap and a black-tipped red bill in summer. It dives for fish and crustacea.

Herring gull, *Larus argentatus*. The quintessential 'seagull', this is a big, bold, noisy bird with a gray back and black and white wingtips.

The royal penguin, *Eudyptes schlegeli*, is a highly social bird. Although flightless, their wings have been adapted to flippers and their webbed feet help propel their sleek bodies through the water.

The fulmar, *Fulmarus glacialis*, is a stocky, gray petrel which feeds largely on dead fish and offal thrown overboard from fishing boats.

billed forms probe the soft sediments in search of worms and mollusks, diving ducks catch small animals in shallow waters, and geese graze on saltmarsh vegetation. Some species, such as the oystercatcher, have developed techniques for feeding on the rich variety of invertebrates on rocky shorelines using their chisel-shaped bill to knock bivalve mollusks from the rock and crack them open.

Detecting food from the air requires high visual acuity. Given the counter shading of many smaller fish, the occasional flash of silver as one fish in the school turns may be the only clue available to a surface flying bird of the availability of food. Penguins appear capable of using a form of sonar, listening to the echoes of the noise made by bubbles collapsing in the wake of fish to detect their prey at depths of several hundred meters. Penguins have become adapted to an aquatic existence so completely that they have lost the power of flight and their wings have become modified to form solid paddles for rapid underwater swimming in pursuit of fish.

Auks are also capable of using their wings in swimming beneath the surface. The wings are held half open, with the primary wing feathers closed, to provide the main source of underwater propulsion. Webbed feet provide steering ability rather than propulsion. This group includes the guillemots, razorbills, auklets and little auks. Several extinct species had adapted to this underwater mode of existence such that they had lost the power of flight. Regrettably, flightless species such as the great auk were extensively hunted for food by sealers, and the great auk became extinct in 1844. These flightless auks may be considered the northern hemisphere equivalent of the southern hemisphere penguins.

The northern gannet, *Morus bassanus*, dives from a great height, plunging into the sea with a great splash, to feed on fish such as mackerel and herring.

he great ormorant, *halacrocorax arbo*, is large bird ith a slender, ook-shaped bill, hich it uses to rasp slippery, uscular fish.

Diving for fish

Diving from the air to catch fish is a mode of feeding which some species, such as the gannets and boobies, have developed to a considerable degree. These birds may plunge from heights of 100 feet (30 meters) or more, so the skull is strengthened to withstand the impact, a layer of fat protects the body, and modified pneumatic air sacs provide a cushion. Many other species plunge, rather than dive. Flying close over the surface, they capture fish just below the surface rather than at depth, while the skimmers actually feed at the surface. In these birds, the lower jaw is greatly enlarged in comparison with the upper and is held just below the surface as the bird flies along, being used as a scoop to capture small fish and invertebrates.

For most areas of the world's oceans, the density of available food resources is comparatively low, and the energy required to collect them is high. As a consequence, many seabirds rear only one or, at the most, two young at a time. Most species are colonial nesters with species such as gannets and fulmars nesting on cliffs, gulls and terns on the ground in sandy areas, and species such as the Manx shearwater in burrows underground. During incubation, one parent shearwater remains in the burrow while the other forages at sea for several days at a time. When they return, each individual locates its own burrow by identifying the call of its mate, landing at night to avoid predation by larger gulls. The fully grown juveniles also emerge at night, making their way downhill to a cliff from which they launch themselves for their first flight. Colonial nesting appears to provide a stimulus for synchrony of breeding and hence improved success. Winter flocks may also improve survival, since a single individual diving for food stimulates the others to follow.

Drinking seawater

An important adaptation of seabirds is their ability to drink seawater. Seabirds have therefore developed specialised salt excreting glands to cope with high salt intakes. These glands lie in a bony socket above the eye and the duct empties through the nostrils. Some species possess a fold of skin with which the nostrils can be closed to prevent seawater from entering when the animals dive. Like most birds, seabirds possess an oil gland at the base of the tail from which oil is squeezed and spread along the feather during preening. The oil prevents the feathers from becoming waterlogged, and diving birds will preen several times a day to maintain the waterproofing of their feathers. Some species, such as cormorants and shags, do not produce sufficient oil to ensure efficient waterproofing and must spread their wings to dry the feathers after diving. Frigate birds, avoid getting wet and snatch fish from the water surface.

▼ ***Red-crested cormorants*** *and horned puffins perch on a rocky outcrop on the Pribilof Islands, Alaska.*

Population numbers

Seabird populations are affected by man's activities, with scavenging birds being positively affected by the increased volumes of offal jettisoned at sea by factory trawlers. Others are adversely affected by reduction in the size of the fish populations on which they feed. On a localized, but nevertheless large scale, populations of seabirds have been dramatically reduced by large-scale oil spills such as those of the *Exxon Valdez* and the slicks resulting from the 1991 Gulf War. In addition, seabird populations which depend on stocks of a single species in zones of upwelling, such as the cormorants, boobies and brown pelicans which feed almost exclusively on the Peruvian anchoveta, have been dramatically reduced by El Niño events, from an estimated level of 30 million in 1950 to about 300,000 in 1983.

► ***The main hunting*** *techniques used by seabirds vary according to species and also the conditions of the sea. Terns make shallow dives; gannets and boobies plunge below the surface from 100 ft (30 m); while auks and penguins pursue fish below the surface. The cormorant uses its feet for swimming, while penguins and auks are much more likely to use their wings.*

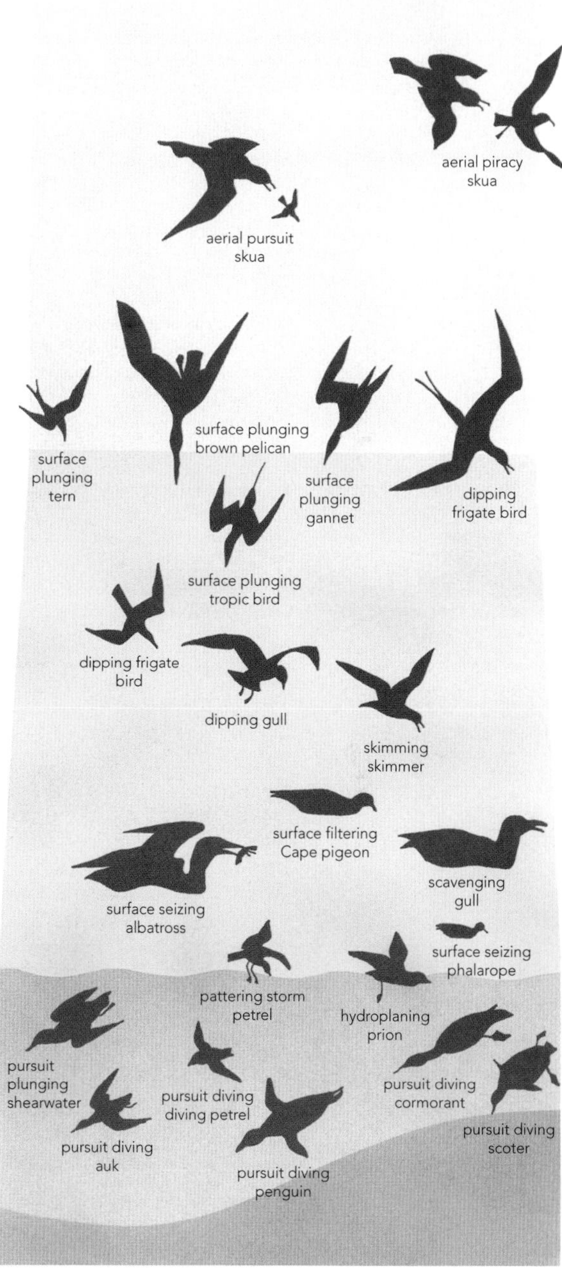

Migration is the term applied to a regular journey made by a particular species of animal, either on an annual or on a lifetime basis. Migratory movements of marine animals are made in response to breeding patterns or to seasonal changes in the availability of food. The long distance migrations of baleen whales across whole ocean basins are made in response to seasonal changes in the availability of their planktonic food, the animals moving into higher latitudes during the summer and into lower latitudes or the opposite hemisphere during the winter. Other migration patterns may involve moving from breeding to feeding grounds. In some cases, the individual animal makes the journey only once, in others the journey may be made year after year.

Among marine mammals and birds, individuals aggregate at suitable locations to court, give birth and care for the young. Such breeding aggregations also occur among many species of fish and may involve movements over relatively short distances, as in the case of reef fish, or over long distances, in the case of tuna, herring and many pelagic species. Some small, coastal wading birds undertake spectacular annual migrations. Knots, for instance, breed in Arctic and subarctic regions before migrating 10,000 miles (16,000 kilometers) or more to wintering areas at the southern end of Africa, South America and Australia. Several species of terns in East Asia arrive in the region of the South China Sea in September and depart for the northern coastal areas of China and Russia in April. The white-winged tern is one such species: arriving in its white winter plumage with a black crown, it departs in summer plumage of black head, body and forewings. Such migrations follow the same, generally coastal routes, and the same staging areas are used each year, reflecting the specific requirements of these species for food resources and feeding grounds in coastal areas.

▼ ***Knots*** *undertake long annual migrations, up to 10,000 miles (16,000 kilometers) from their breeding grounds in the Arctic to the southern hemisphere during the Arctic winter.*

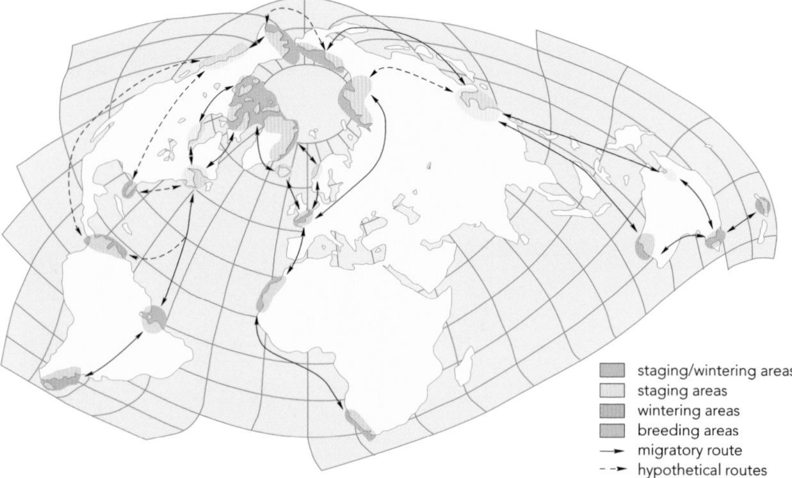

▲ ***A female** Leatherback turtle,* Cermochelys coriacea, *in Grande Riviera, Trinidad, returns to the sea at dawn following egg laying. Marine turtles mate in the ocean, after which the females come ashore on sandy beaches to dig a nest, into which they lay their eggs. The eggs incubate beneath the sand until they hatch, whereupon the young turtles make a perilous journey back to the sea.*

Breeding migrations

For most marine animals, which are members of the pelagic community, migratory patterns are an integral part of their life cycle. Spawning in many open-ocean fish species takes place where currents will carry the larvae in a planktonic community that provide the young animals with food. Thus tuna migrate annually across the Pacific and Atlantic basins to spawn in specific areas before continuing on to adult feeding grounds. The juvenile tuna float with the plankton to nursery areas and ultimately join the adults in the feeding grounds. In the case of turtles, migration is again associated with reproduction, individual female turtles returning to nest on the beaches on which they were hatched. Suitable nesting beaches seem to be the main determining factor in the egg-laying migrations of these animals, as is the case for many seabirds which return to nest on the same cliff year after year. Green turtles swim the entire length of the South China Sea to return to their feeding grounds in North Borneo after laying eggs in Pulau Redang on the east coast of peninsular Malaysia.

Some breeding migrations, such as that of the gray whale, are not determined by the requirements of the young, which remain in the company of the parents throughout life. This species makes an annual migration of some 5000 miles (8000 kilometers) from the Bering Sea in the North Pacific to the warmer waters off Baja California to breed.

A large number of different fishes migrate between freshwater and saltwater, of which the various species of salmon in the northern hemisphere are well-known examples. Spawning takes place in freshwater, usually some distance inland where the rivers are shallow and well oxygenated.

Eggs are laid in gravel banks, and the larval fish spend a long time in the freshwater reaches of the river, gradually moving downstream during their development to spend their subadult life at sea. On reaching sexual maturity, they return to the rivers from which they were spawned, and the cycle repeats itself. In the case of the sockeye salmon of North America, the adults die immediately after spawning.

In the case of the North Atlantic salmon, however, individuals survive, returning to the sea after spawning; they may make two or three migrations in a lifetime. It has been shown that salmon recognize the river in which they were spawned by subtle chemical cues, which distinguish water from one river to the next. Only successful spawning results in individuals completing the cycle. When adults enter an unsuitable river and spawn, few or no individuals will survive to return and breed. Restocking rivers, restored following degradation

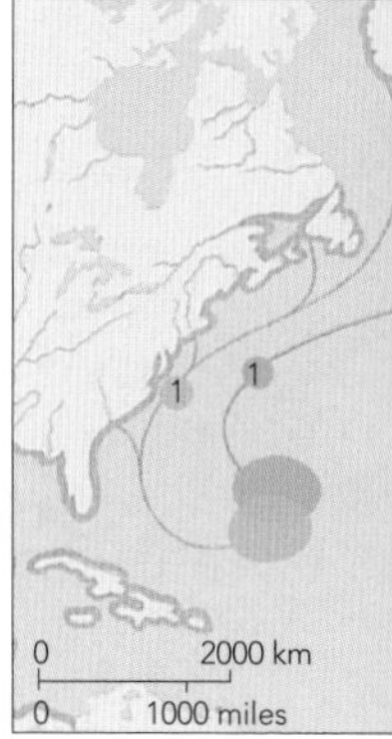

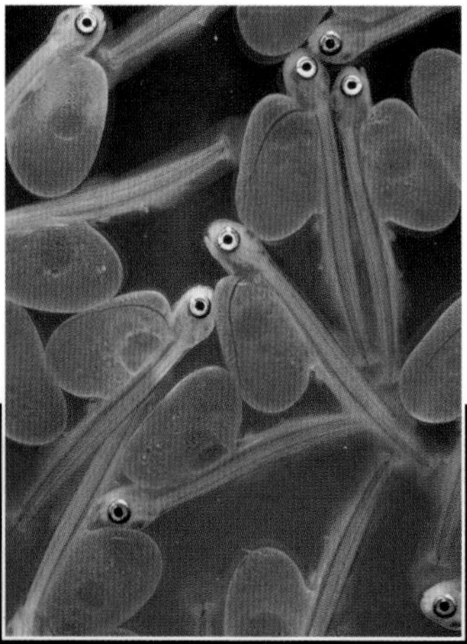

◄ **Newly hatched salmon alevins** *Many salmon return to the stream or lake where they were born in order to spawn. Only about 2 percent of all salmon hatched survive to adulthood.*

▼ **Migrating adult** *American eels rest together underwater, near Maryland, USA. These freshwater eels migrate to the ocean to spawn upon reaching maturity.*

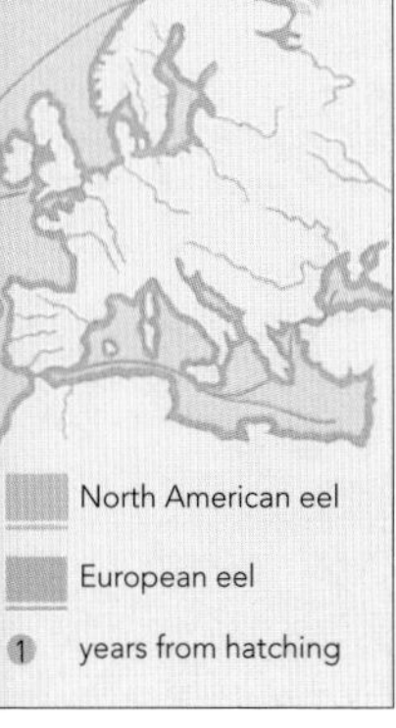

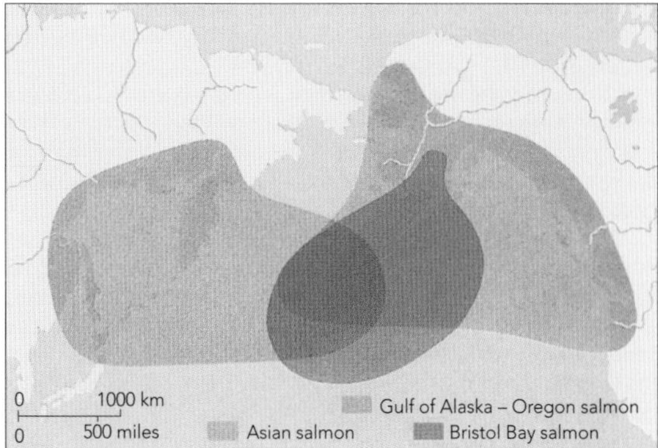

▲ ***Both the North American** and European freshwater eels spawn in the Sargasso Sea (left). The larval eels are carried by the Gulf Stream toward North America and Europe, where they enter streams and rivers and spend their life feeding and growing. On reaching maturity some years later, the eels return to the Sargasso Sea to spawn and die. Unlike the eels, salmon (right) spend most of their life at sea, returning to freshwater on reaching sexual maturity to spawn some distance inland. Atlantic salmon may make two or three spawning runs in a lifetime, while Pacific salmon, which divide into three distinct stocks, spawn only once, and die soon after.*

from pollution, requires that the juvenile migratory species be exposed to water from that river so that they can identify it and return there to spawn.

Atlantic crossing

In some species, the process is reversed; thus freshwater eels from North America and Europe move to the Sargasso Sea to spawn. The larval eels are then carried by currents northward along the North American seaboard and across the Atlantic to Western Europe. Here the young elvers enter rivers, moving upstream to spend their life feeding and growing in the freshwater environment before reaching sexual maturity and undertaking the lengthy migration back across the Atlantic to the spawning grounds of the Sargasso Sea.

Although few invertebrates undertake true migrations, since most larvae drift passively with the ocean currents, settling out in any suitable area when their larval life is completed, there are some spectacular exceptions. Many tropical land crabs which live throughout the year inland, away from the marine environment, migrate annually in enormous numbers to shed their eggs into the sea, while tropical spiny lobsters migrate inshore to spawn, often forming lines or columns of marching individuals across the sea floor. Once inshore, the eggs are released and the larval lobsters are generally carried passively offshore by the prevailing currents before settling on the bottom.

In tropical coastal environments, mangroves and seagrass beds form important nursery areas for many species of fish and shrimps; the complex fronds and root systems provide a multitude of hiding places. Many species of smaller, tunalike fishes, which spend their adult lives offshore, will come inshore to spawn in close proximity to mangrove systems. Here the larval fishes hatch before moving gradually offshore. Penaeid shrimps, which inhabit soft-bottom areas of deeper water, also use mangrove areas as nurseries where the juveniles feed and develop during the early stages of life before migrating offshore to deeper water to complete their growth, returning inshore on reaching sexual maturity.

In the pelagic environment, vertical grades, including those of light, temperature and salinity, determine the distribution of organisms. The greatest changes are seen at the boundary between the mixed surface waters and the deeper water masses. In some shallow-water areas, a thermocline, or boundary, develops between the warmer surface and the cold water below. Across such thermoclines, temperatures may drop rapidly.

The epipelagic region, or zone of light penetration, stretches from the surface to a depth of around 650 feet (200 meters) and is followed by the mesopelagic zone between 650 and 6500 feet (200 and 2000 meters). Below this the bathypelagic zone extends from 6500–21,000 feet (2000 to 6000 meters) in depth, while the deepest ocean trenches contain an abyssopelagic zone greater than 20,000 feet (6000 meters) deep.

Each of these zones contains a different community of species adapted to these different conditions of salinity, temperature and light intensity. The differences are greater in the bottom-dwelling, or benthic, community than in the pelagic community, where many deep-water species range over considerable depths. Environmental conditions are more stable at greater depth in the ocean than they are at the surface, temperatures vary little over great distances compared with the surface, and in some instances, species of animal found at or near the surface in polar seas also occur in tropical ocean areas at depth, where water temperatures are cool.

At depth, the strength of water currents is low, and from the perspective of animals living there, can be considered negligible. Although nutrients tend to be at higher concentrations at greater depth, this has little impact on the organisms, since they are only used directly by plants in the epipelagic zone. Except for isolated communities based on the chemical energy supplied by thermal vents, the entire community below the epipelagic zone depends on the rain of fecal materials and dead organisms, which sinks down from the epipelagic zone. Deep-sea thermal vents are home to their own specialized community of crabs, other crustacea, mollusks and tubeworms that may reach a meter in length.

The epipelagic zone contains the most abundant life, which depends, both directly and indi-

▼ ***An imaginary profile*** *of the typical coastal and oceanic zones, with a selection of the life-forms that might be seen off the Pacific coast of Central America. The animals illustrated are not drawn to scale. Tiny plant and animal plankton, the basis of life in the ocean, occur in great quantities: their presence has been indicated and examples of the major types illustrated. The density of life is very high in the upper sunlit zone so, to show a reasonable selection of the vast numbers of inhabitants, the depth of the body of the diagram has been distorted.*

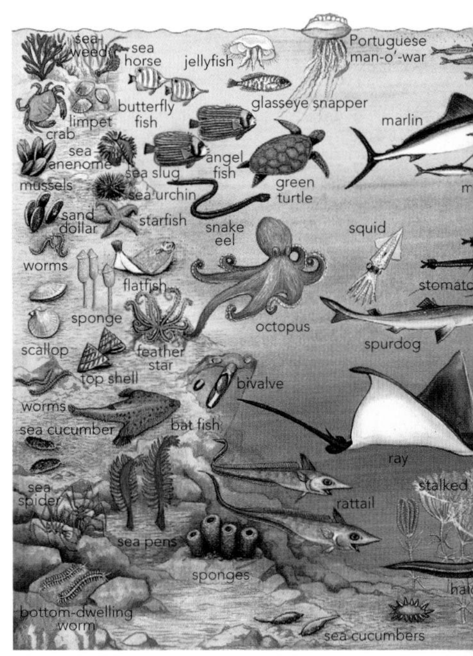

rectly, on the primary production of the plankton community. Passing down through this zone, the proportion of carnivores increases in the pelagic community as the density of primary producers declines. In addition, eye size increases as the light intensity decreases. With the decrease in light, colors change, from blue at the very surface through transparent in the upper layers, to red and black at depths of 330 feet (100 meters) or so. The surface waters down to depths of around 500 feet (150 meters) are dominated by swift predatory fishes, such as tunas and swordfishes, together with swarms of smaller species, such as the lantern fishes.

The abyss

In the aphotic, or abyssal, depths the only light available is that produced through bioluminescence – light signals produced by animals, such as anglerfish to lure prey within reach or by the lantern fishes to signal between members of the same species. Many of the bioluminescent organs of deep-sea animals rely on the presence of symbiotic bacteria contained in these special structures to produce the flashes or continuous beams of light.

sunlit zone
650 ft (200 m)

twilight zone
3000 ft (1000 m)

dark zone
19,500 ft (6000 m)

trench zone
33,000 ft (10,000 m)

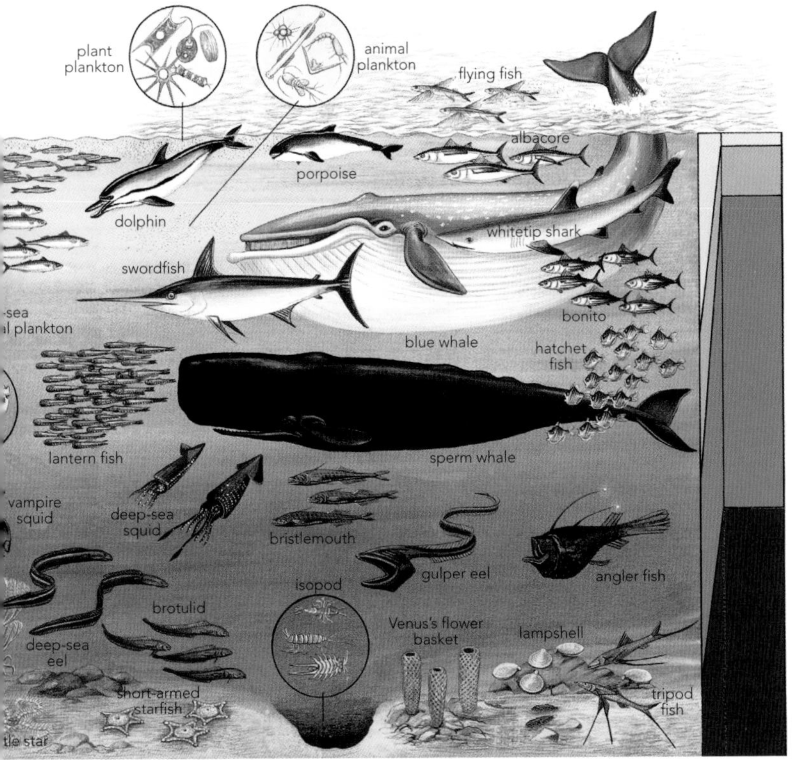

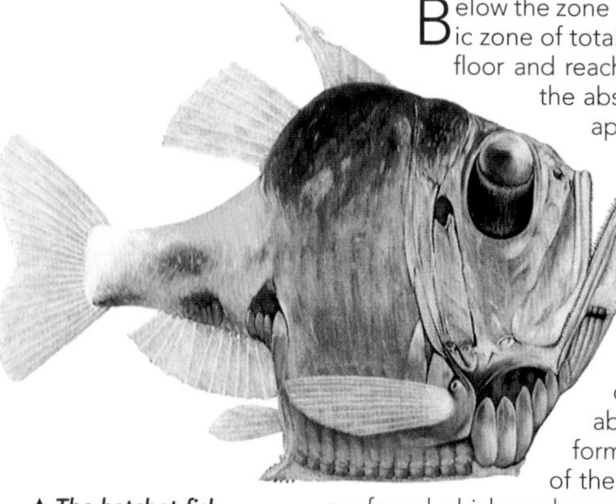

▲ ***The hatchet fish**, Argyropelecus, grows to about 4 in (10 cm). It has huge jaws and is capable of swallowing quite large prey, despite its small size and flattened body. It must rely on occasional meals due to the very low density of prey in the ocean depths.*

Below the zone of light penetration lies the aphotic zone of total darkness, extending to the ocean floor and reaching almost from pole to pole. In the absence of light, most species in the aphotic zone are a uniform brown, black or violet and generally quite dull.

Within this region, conditions vary little over great distances; temperature varies little with depth; day and night and seasonal changes are not apparent; water circulation and current speeds are trivial; and the abyssal plain is covered with a uniform layer of fine ooze. In the regions of the midoceanic ridges, thermal vents are found which are home to curious communities of organisms, including meter-long tube worms, filter-feeding bivalves and crustacea that depend on the presence of chemotrophic bacteria. These bacteria use sulfur from the emissions of the vents as a source of energy, thus providing a food source to support the entire community.

Given the uniform conditions found at depths in the ocean, it is not surprising that the fauna shows little change over great distances. Of the 11 species of the upper continental slope community found at a depth of 1500 feet (450 meters) off the west coast of Africa, no fewer than eight are important members of the same community off western France. The environments in these two areas differ by less than 1°C in average temperature, although they are separated by 40° of latitude.

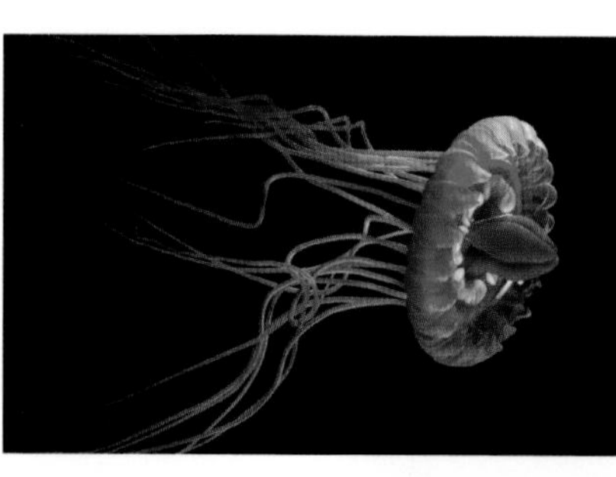

Declining food resources

Below the lit zone, food supplies generally decrease with depth, meals are infrequent and animals found in the mid waters and abyssal depths of the oceans have reduced skeletons, mus-

▲▶ **Atolla *jellyfish*** *are common in the deep ocean. If disturbed, the Atolla creates a circular wave of luminescent light around its outer edge. Harbor Branch Oceanographic Institution in Florida have created a camera that mimics the jellyfish's bioluminescent display.*

cles and other body tissues, reflecting the low availability of food. Between 3200 and 6400 feet (1000 and 2000 meters), many fish lack swim bladders, maintaining neutral buoyancy by other means, such as the retention of the chemical ammonia.

Many deep-sea fishes possess large heads and mouths, and long thin bodies with an extraordinary capacity to expand. The great swallower, *Chiasmodon niger*, can consume prey almost the same size as itself, while in the gulper eels and ratfishes, the head is huge and the remainder of the body is reduced to a long, almost whiplike, tail. Although sparse, the abyssal fauna is highly diverse, the pelagic community of abyssal fishes numbering some 2000 species in contrast to the 200 or so inhabiting the upper reaches of the ocean.

Filter-feeding animals decline in abundance with depth, while animals which ingest sediments continue to occur even at the greatest depths. Burrowing animals predominate on mud bottoms, but on the softer, unconsolidated sediments of the deep ocean floor, a variety of bizarre species occur, such as the tripod fish, with enlarged spines that support the body above the surface of the sediment. Sea pens and glass sponges are supported upright in the sediments, filtering particles from the water column, while sea cucumbers move across the surface of the sediment.

As a result of the low temperature of deep waters, between 35°F and 41°F (2°C and 5°C), the species have low metabolic rates, slow growth rates and live for a long time. Bivalve mollusks at 10,000 feet (3000 meters) off the east coast of North America are up to 250 years old, although only 1 inch (2.5 centimeters) in length.

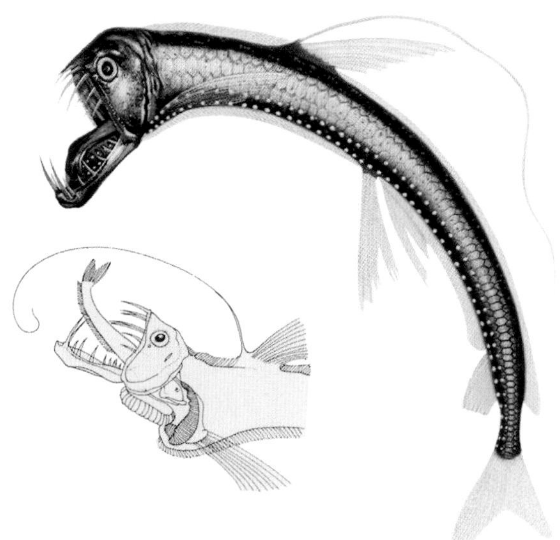

► ***Many deep-sea fishes*** *possess large heads and mouths, and long thin bodies capable of great expansion. The viper fish (right),* Chauliodus *sp., has huge recurved teeth and widely gaping jaws, which enable it to catch prey much larger than itself. It attracts prey by means of a light organ on the end of the first ray of its dorsal fin, and when it moves in to attack, it throws back, the lower jaw shoots forward and the prey is swallowed and digested extremely rapidly.*

▲ ***The limpet** clamps to the rock surface using its muscular adhesive foot.*

▲ ***Mussels** live in dense colonies, attached to rocks by byssus threads.*

▲ ***The cockle** filter feeds on plankton at high tide using its two siphons.*

▲ ***The fiddler crab** is a familiar inhabitant of tropical muddy shores.*

The intertidal zone is one of the harshest environments on Earth with conditions across a shoreline following a grade, from purely terrestrial above the zone of influence of salt spray, to the purely marine below the level of the lowest low tide. Animals and plants living here must be able to withstand twice daily inundation by the tides.

Intertidal animals and plants therefore display numerous adaptations to resist desiccation during low tide, including remaining tightly clamped to the surface of rocks, as in the case of limpets; closing the shells with a tight-fitting operculum, as in the case of snails; or retreating to tidal pools, crevices, burrows and beneath seaweeds to avoid the drying action of the wind and the direct heat of the Sun. As the tide returns, the organisms must be capable of withstanding the battering of the waves. Seaweeds have developed strong holdfasts to remain attached, while animals such as mussels have developed byssus threads. Others, such as oysters and barnacles, cement themselves to the surface of the rocks. At high tide, predators such as fish move in to feed on the soft-bodied animals, which become active under water. When the tide is out, the shore is a feeding ground for birds.

On soft shores, much of the fauna lives in the sediment. Burrowing requires not only the construction of a suitable burrow, but also some mechanism for collecting food. Many burrowing animals extend a crown of tentacles above the surface to feed on particles in the water column, while others plow through the sediment, consuming it like earthworms.

Distinct zones

Most shoreline organisms occur in recognizable zones, reflecting the grade of environmental conditions across the shore. Yellow, white and gray lichens occur in the splash zone and green seaweeds in the upper intertidal. Below the zone of green seaweeds are found the brown species, including the characteristic bladder wracks (*Fucus spp.*), plants with gas-filled bladders, which float when the tide is in and keep the fronds apart. Below the brown seaweed zone occur the red seaweeds, while in the subtidal are found the large kelps of genera such as *Laminaria* which grow to a large size at rates of up to 3 feet (1 meter) a day.

◀ ***Tide pool** at West Quoddy Head, Maine, northeastern United States. On full and new moons (spring tides), tides at the adjacent Lubec Embayment exceed 20 feet (6 meters), the greatest tidal range on the east coast of the United States. Seals are often seen swimming in the tides.*

Competition for space between sessile species is important in determining the zonation of species; the distribution of the two barnacles, *Balanus balanoides* and *Chthamalus stellatus*, illustrates this. The latter species is found in the top third of temperate rocky shores, while *Balanus*, which is less resistant to desiccation, occurs lower down. Individuals of this species die from desiccation, while in the lower third of the intertidal they are preyed upon by a predatory gastropod, *Thais*, which is itself limited to the lower third of the intertidal.

Chthamalus is, however, more resistant to desiccation, and larvae which settle high on the beach outside the range of *Balanus,* grow and have high survival rates; those which settle lower on the shore, within the range of *Balanus*, suffer high mortalities, since the more upright shell of *Cthalamus* is undercut by the flatter profile *Balanus*. As *Cthalamus* grows, the shells of *Balanus* grow under it and lever it away from the surface.

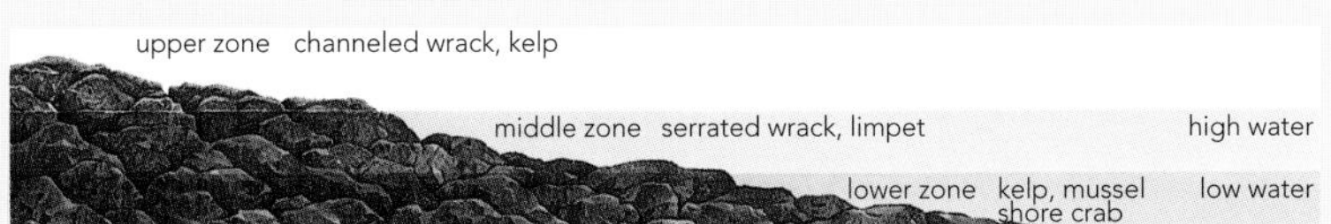

Rocky shores support a wide variety of animal and plant species which are zoned in relation to the tidal height. Green seaweeds are found on the upper shore, brown and red lower down, with the large kelps below low tide. They provide food and shelter for a range of animals which hide beneath them at low tide. Rock pools harbor a large variety of animals, some of which cannot survive the dryness of the beach at low tide.

The sandy shore has a surface which is constantly moving under the action of the tides, so it provides no place for attachment and no inviting crevices, but can hold water between its minute particles. Beneath the surface, the environment is unaffected by the weather. A wide range of animals inhabit the space between the sediment particles at low tide, while others burrow and filter food from the water at high tide.

Muddy shores have low water movements and are rich in organic matter, often supporting growths of seagrasses, such as Zostera, and algae, that provide a plentiful food supply for grazing animals. As the particles of muddy shores are so fine there is little space between them and oxygen is at low concentration. Burrowing animals, such as clams, may extend siphons above the surface down which a current of water is drawn.

Coral reefs, which are a dominant feature of tropical shore lines, have been referred to as the 'rain forests of the sea'; and like rain forests, they are extremely diverse, being home to an enormous variety of animals and plants. Although they are generally found in areas of low marine productivity, reefs themselves are highly productive. The high productivity is due to the reefs' efficient cycling and reuse of nutrients, which are in short supply in the surrounding oceanic waters.

The primary production of coral reefs is between 30 and 250 times as great as that of the open ocean, and coral reefs may produce 44–146 ounces per square yard (1500–5000 grams of carbon per square meter) per year.

How reefs are formed

The living reef is composed of a thin veneer of living coral colonies growing on the surface of older, dead coral skeletons. Coral colonies that die or are broken off during storms, break down to form sand, which fills the spaces between the frame-building corals. If the land on which the reef is growing sinks, or sea level rises, as is happening at the present time, the reef continues its upward growth. Eventually, with the passage of time, the living reef community may be growing on many hundreds of meters of solid coral rock. The living reef at Enewetak atoll in the Marshall Islands, for example, grows on a base of 4500 feet (1370 meters) of coral limestone. This limestone has been built up on top of a volcanic cone that rises some 16,400 feet (5000 meters) above the ocean floor.

Reef-building corals do not grow well below 70–100 feet (20–30 meters), since they contain microscopic algae, or zooxanthellae, which require sunlight for photosynthesis. The remarkable association between the coral animal and

▼ ***Coral reefs*** *grow in warm waters over 68°F (20°C), thriving at about 75°F (24°C). Reefs grow beyond the tropics in sufficiently warm currents. Corals do not grow in freshwater discharges and cannot grow in water full of sediment, which smothers the fragile reef organisms. Although the optimum depth for growth is a few meters below the surface, where oxygen and sunlight are abundant, coral can grow at up to around 132ft (40 m). In ideal conditions, healthy reefs grow up to 1in (25mm) in a year.*

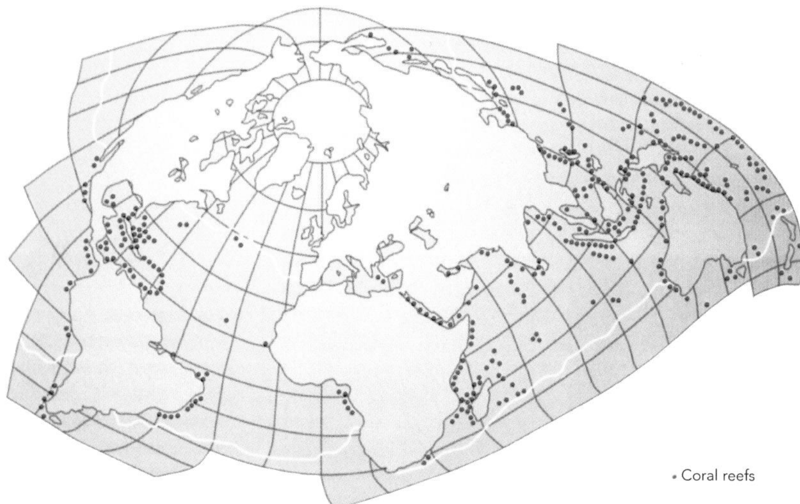

fringing reef

barrier reef

coral atoll

seamount

▲ ***An atoll** starts to form as soon as a volcano rises above the sea. Corals colonize and form a fringing reef around the island. The reef grows upward as the island sinks, forming a barrier reef enclosing a shallow lagoon. Eventually, the island disappears leaving a ring-shaped reef or atoll. If subsidence occurs faster than coral growth, then the reef dies leaving a seamount below the surface of the sea.*

the zooxanthellae benefits both partners, with the algae deriving nutrients and carbon dioxide from the coral while speeding up the rate of skeleton formation of the coral. Other algae, both unicellular and macroscopic, are important to the productivity of reefs, while encrusting, calcareous forms cement loose material together, thus stabilizing the surface of the reef for the colonization of larval corals.

Solitary corals are widely distributed, but do not form reefs, which are best developed in the tropics and subtropics where water temperatures range between 68°F and 86°F (20°C and 30°C). Although reefs grow well at 64°F (18°C) in the Florida Keys and above 91°F (33°C) in the Northern Great Barrier Reef of Australia and the Persian Gulf, most reefs grow in areas with water temperatures of around 75°F (24°C).

The coral animal

Corals are colonies of tiny individual animals or polyps that secrete a skeleton of calcium carbonate. Colonies grow by adding new individuals, and the shape of the colony ranges from the compact brain corals found in areas of high wave energy, through heavy branching and plate corals in deeper water off the reef edge, to smaller, finely branched forms found in more sheltered water behind the reef crest. The most active coral growth is generally on the outer edge of the reef. Here, water movement is greatest, carrying with it plankton on which the coral polyps feed.

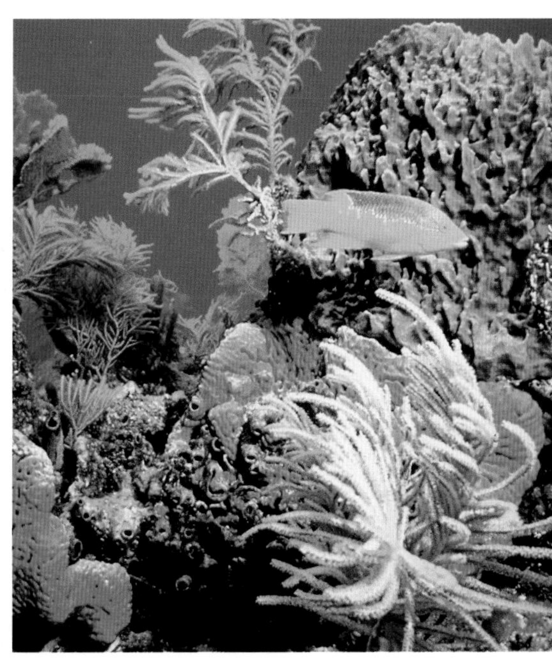

▶ ***Coral reefs** are home to an amazing variety of spectacularly colored fish. The coral provides the fish with numerous hiding places and abundant food supplies.*

In addition to reproducing by simple budding, so increasing the size of the colony, corals also reproduce sexually. At a specific time each year and over a period of only 3 to 5 nights, corals release millions of eggs and clouds of sperm, turning the sea a luminescent green and providing a nutritious 'soup' for other reef organisms. The surviving eggs hatch as larvae that settle out to form new colonies.

Coral reef diversity

While it is true that coral reefs are among the most diverse marine habitats in the world, their diversity is much less than that of rain forests, which have more than 250 species of tree per hectare. A single, large reef system may support around 200 coral species, and only about 1000 species of reef-building corals have been described worldwide.

The center of coral diversity is the southeast Asian region, with some 700 species being found in the Indowest Pacific, compared with only around 35 in the Atlantic. More than 400 species of hard corals are believed to occur in the Philippines, and as one moves away from this center of diversity, the numbers of species decline.

The wide diversity in growth forms of different corals provides a multitude of microhabitats, refuges and food sources for other organisms, the diversity of which is also high. Coral reefs are believed to support a third of all living fish species, for example. While some animals, such as the crown of thorns starfish (*Acanthaster plancii*) and the parrot fishes (*Scarus* spp.), feed on corals directly, many other animals use the reef as a place of attachment. Sea fans, feather stars and

▼ ***The Pacific island*** *of Bora Bora is an extinct volcano that is gradually subsiding. In time, the inner island will disappear and a true atoll will be formed*

Coral formation

Coral colonies consist of large numbers of polyps, which secrete a skeleton of calcium carbonate. Individual polyps secrete their own theca or cup, into which they can withdraw. Each individual is connected to its neighbor by tissue.

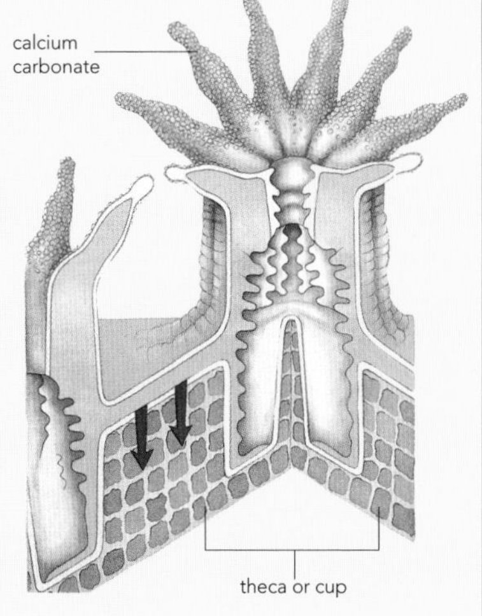

sponges grow attached to the surface and filter feed on the plankton and suspended matter contained in the surrounding water. Up to 80 percent of the plankton in the water passing over a reef, may be removed by corals and by other filter-feeding organisms.

In sandy areas, sea cucumbers feed on surface detritus or eat the sand directly, digesting the bacteria and micro-organisms which grow on dead organic matter. Grazing animals, such as sea urchins and many mollusks, feed on the film of algae, which grows on all dead surfaces of a reef, while small fishes crop the fine algal turf and are themselves eaten by predators, such as moray eels. The moray eel is not immune from predation, being eaten by the venomous sea snake *Laticauda colubrina*. To avoid predation, many reef animals are brightly colored, advertising the fact that they are distasteful or poisonous. The bright colors of many other active reef animals serve, however, as signals, enabling individuals to recognize members of their own species of the same or opposite sex.

Present threats

Unfortunately, the diverse and beautiful world of the coral reef is as threatened as the tropical rain forests. It has been estimated that as much as 10 percent of the world's coral reefs have been degraded beyond recovery by human activity. A further 30 percent is likely to decline seriously within the next 20 years, and it has been suggested that more than two-thirds of all reefs may collapse ecologically within the next 80 to 100 years. Such a gloomy picture is not universal; recognizing that much of their tourist income depends on the health of the reef systems, many developing country governments are taking active steps to protect and maintain their living reefs.

Ocean Resources

FISHING

Fish, both shellfish and finfish, are a vitally important source of food for people living around the world's coasts. In 1985, the FAO (Food and Agricultural Organization) of the United Nations estimated that for 60 percent of the population in tropical developing countries, between 40 and 100 percent of their animal protein came from fish.

In temperate regions, most fisheries rely on large stocks of single species which are fished commercially using modern technology. At lower latitudes and in inshore areas, the wider diversity of fish is reflected in the range of techniques used to catch the different species. As much as 75 percent of the world's harvest of fish comes from within 6 miles (9 kilometers) of the shore. The artisanal fisherman, therefore, is a major producer of edible marine products around the world.

The diversity of techniques employed to harvest fish is enormous. Spears, and bows and arrows may be used to catch single fish in shallow water, while hook-and-line techniques are used on set lines, with rods, or trolled behind a boat. Other methods of catching fish involve the use of nets

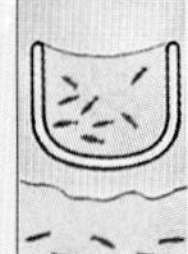

Rock-wall traps *were built on gently sloping shores between high tide and low water to strand fish as the tide receded. This simple trap has been used since prehistoric times in the Pacific.*

Fish species *that swim parallel to the coastline may be deflected into nets by strategically placed fences. The seaward end of the fence leads to a heart-shaped trap.*

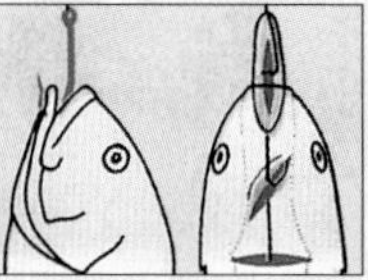

Barbed hooks *and gorges made of bone or wood are used to catch fish which strike a baited line. The simple gorge is hidden in food, and rotates and lodges in the gullet.*

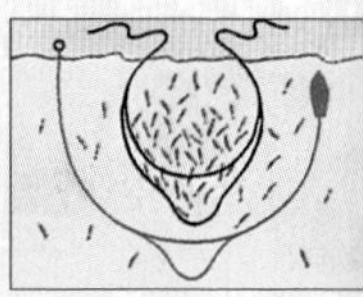

Small shoals *swimming close to shore may be encircled with a beach seine paid out from a small boat. Fish are frightened into the central bag by shouting and splashing.*

In shallow muddy water*, or where the seabed is too rough for nets, fish are often caught using plunge baskets. Fishermen work slowly forward, trapping the fish under hand-held pots.*

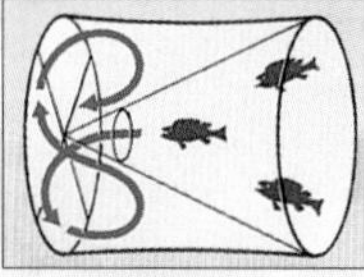

Fyke nets *are a series of conical nets leading from one to another and ending in a cylindrical chamber. Wings of netting fan out from the entrance into the current to increase the catch.*

In shallow water*, where the seabed is smooth, a net may be skimmed across the bottom. The rounded ends of the frame ensure that the net glides easily over the ocean floor.*

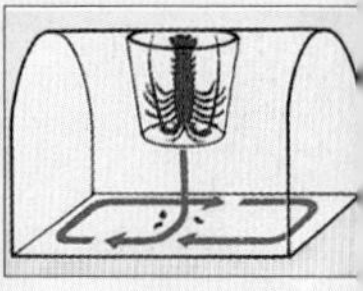

Lobster pots*, usually made from heavy netting or basketwork on a wooden frame, are generally baited with fish offal. The darker they are the better, as lobsters like to hide in the dark.*

► ***Fishermen** on Ifaluk Atoll in the Caroline Islands, Micronesia, return from a fishing trip with their catch of tuna. Their vessel is a traditional outrigger sailing canoe.*

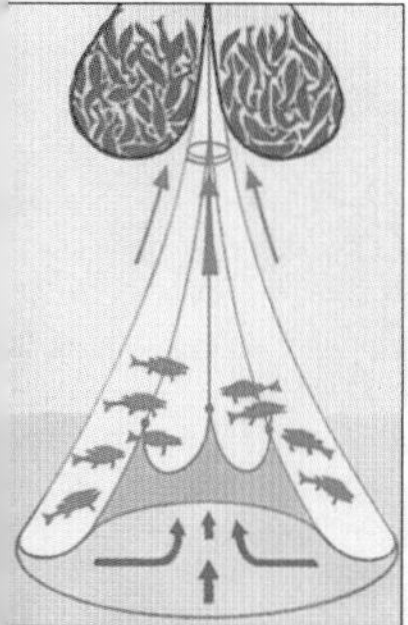

***Throwing a cast net** requires great skill. The weighted edge pulls the net down over the fish, pinning them to the bottom. The net lines are then hauled in, slowly closing the net around the catch.*

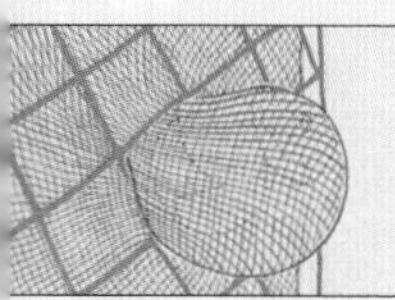

***The trammel net** consists of two coarse outer nets either side of a fine entangling net. Although it is bulky and conspicuous, the fish are blindly driven into the net by people splashing in the water.*

ranging from hand-held, frame nets operated by a single fisherman in shallow water, to large nets that require several men to set and haul.

Gorges, hooks and lures

The simplest fish-catching device attached to a line is the gorge. Made from wood or bone, the gorge is pointed at both ends and when swallowed by a fish it lodges in the gullet. Hooks are more sophisticated and come in a variety of shapes and sizes, depending on their place of origin and the target fish species. The shank, for example, may be straight or curved, and the point barbed or plain. Many hooks are small, having been ground from a single mollusk shell Others, such as those used to catch the deep-sea oilfish, *Ruvettus*, are largely made from shaped wood and shell. Trolling, or dragging, a lure behind a boat catches fish that mistake the lure for potential prey. Traditional lures of shell, bone and feathers were once widespread among the fishing communities of the Pacific, but today have been replaced by plastic lures and steel hooks.

Traps and nets

An ancient element of the fishing technology of southeast Asia is the thorn-lined trap constructed of spiny branches tied together in a cone, with the spines pointing toward the tip of the cone where the bait is secured. When the fish enters, the spines lodge behind the gill covers, preventing the fish from wriggling out backward. Larger traps include fish fences constructed in mangrove and estuarine areas, weirs in river channels, and stone traps on reef flat areas. Such traps rely on catching fish as they move into and out of feeding areas with the tide.

Traditionally, nets, both large and small, were made of vegetable fibers. Recently, however, nylon has replaced the original materials. Small, hand-held nets are used to dip fish from the water, and larger nets are set as temporary traps, catching fish by entangling them. The circular cast net is a widespread element of fishing technology; its outer fringe is weighted, and when the net is cast, the weighted rim sinks faster than the center, trapping fish inside.

Artisanal fishermen generally operate singly or in small groups on small vessels, while commercial and industrial-scale fishing is based on much larger vessels that operate at considerable distances from their home ports. Long-distance fishing fleets not only catch fish, but also process the catch on board. Commercial fisheries account for about 90 percent of the total world fish catch, of which 75 percent is for human consumption and 25 percent is used to produce fish meal for animal feed.

Development of commercial fisheries

Large-scale commercial fishing began in the North Sea as early as the 15th and 16th centuries, and by the 17th century, fishermen from Devon had already crossed the Atlantic to fish for cod off the coast of Newfoundland, Canada. The introduction of steam and then diesel engines fueled a rapid expansion of long-distance fishing. Not only could such vessels travel further and faster, but also by carrying ice, they could extend the time that fish could be held without salting. This led to a rapid increase in fish catch in the North Atlantic, and fishing vessels were soon penetrating the Arctic Circle in search of cod and haddock.

Following World War II, improved refrigeration technology enabled vessels to remain at sea until their holds were filled, while improved catching technology and the increasing size of vessels also resulted in the exploitation of new stocks, and commercial fishing extended outward from the North Atlantic to other areas.

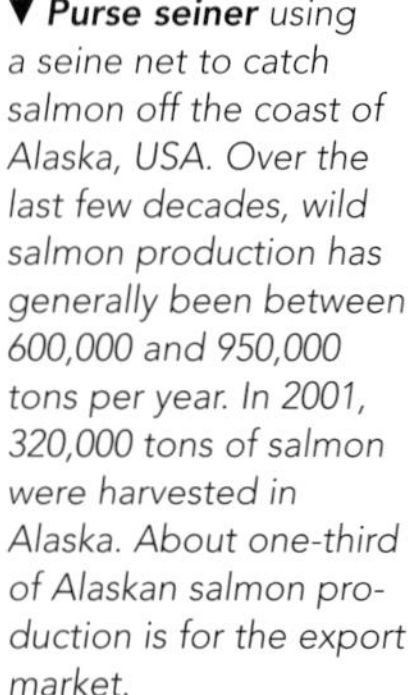

▼ ***Purse seiner*** *using a seine net to catch salmon off the coast of Alaska, USA. Over the last few decades, wild salmon production has generally been between 600,000 and 950,000 tons per year. In 2001, 320,000 tons of salmon were harvested in Alaska. About one-third of Alaskan salmon production is for the export market.*

► ***Fishermen*** *shovel pollack on the deck of the Saga Sea, a factory-trawler fishing in the Bering Sea, Alaska.*

▼ ***School of mackerel*** *detected by sonar: notice the shadow of the school on the seafloor (far right).*

Large vessels

Giant purse seiners evolved from small Norwegian mackerel and herring ring netters and were developed for fishing schooling, pelagic fishes, including sardines, anchovies and tuna. The large-scale American vessels of this type are capable of encircling an entire tuna school and holding approximately 2000 tons of frozen fish.

Oceanic long-liners were developed to fish the deeper-swimming tunas and use, as their name suggests, long lines of up to 20 miles (30 kilometers) in length fit with multiple hooks. These vessels range the entire extent of the tropical oceans in search of tuna, marlin and billfish. Drift nets operated by some fleets may be tens of kilometers in length, and are destructive, catching a number of nontarget fish, marine mammals and turtles. They are banned under a regional convention in the South Pacific.

Factory trawlers are large-scale vessels developed to exploit demersal fish and used on the continental shelf areas by long-distance fishing fleets from the former Soviet Union and many Asian states. Large trawl fleets operate in offshore tropical areas where the target is frequently penaeid prawns and tropical spiny lobsters, the stocks of which have been dramatically overharvested in many areas.

By-catch

The by-catch, or nonedible species, of many of the commercial fisheries is a major source of concern to environmental groups and development economists in many tropical countries. In particular, large-scale trawl operations for penaeid shrimps and tropical lobsters not only damage the benthic community, but result in a large catch of fish which are normally thrown back in order to retain freezer space for the more valuable prawns. Such by-catch is normally dead when returned and could alternatively be processed for animal feed, or in some instances, sold in those developing countries where the prawn trawling occurs and where protein shortages are frequent.

Although living marine resources currently provide somewhere between 5 and 10 percent of total world food, they supply between 10 and 20 percent of the world's animal protein. It has been estimated that the sustainable harvest of marine living resources is probably around 90 to 100 million metric tons a year. This level had been reached by the late 1980s, and it is believed that 90–95 percent of the world's fish stocks are now fished at, or even beyond, the maximum sustainable levels.

Commercial fisheries

The commercial catch is made up of around 35 fish species of which six – pollack, mackerel, herring, sardine, cod and anchovy – make up more than half the total landings. The demersal catch, which includes prawns and shrimp, is dominated by cod, hake and haddock. These species amounted to around 30 percent of the total world catch of finfish in the early 1990s. The pelagic catch is dominated by herrings, anchovies and sardines, and several of these large, pelagic fisheries contribute significantly to the total world catch. Fluctuations in a single stock can therefore significantly affect total world landings.

The processes controlling the size of fish populations and patterns of recruitment are not well known. The recovery of North Sea fish stocks during World War II led to the assumption that stocks were largely regulated by fishing effort. More recently it has been realized that interannual variations in ocean characteristics, including variations in current patterns, upwelling and salinity, can affect the survival of juvenile fry and hence stock size.

▼ ***While most nations*** *observe the moratorium on commerical whaling, some hunts continue. This photo was taken at Hvalfjorden, Iceland. In 2003, Iceland resumed whaling after observing the ban for 14 years.*

▲ ***Tall ships** and fishing boats fill the harbor at Oban, western Scotland, in this 19th-century photo. In 2003, a man-made reef was created to encourage fish to feed and breed, replenishing declining fish stocks.*

Sustainable levels of harvest

Following World War II, and during the period up to the 1970s, the world fish catch grew at around 7 percent per annum. By the 1960s, when the world fish catch had reached 50 million metric tonnes, it was already understood that some of the stocks were being fished at unsustainable levels. The decline in stocks of cod and haddock in the North Atlantic, herring in the North Sea and salmon in the North Pacific, all attest to the destructive nature of modern harvesting techniques.

As stocks declined, the fishing industry developed more sophisticated techniques for locating and catching fish. At the same time, commercial fishing turned to less desirable species, such as squid and shellfish, and the focus of fishing shifted from the northern to the southern hemisphere.

The Californian sardine fishery of the 1930s is a classic example of the boom and bust cycle that characterizes many fisheries. Fishing boats and gear were poured into this fishery until it became overcapitalized, with too many boats chasing a diminishing supply of fish. The stock finally crashed to extinction in the early 1950s.

A need for regulation

In the past, when the numbers of fishermen were low and fishing technology was less sophisticated, the concept of open access to fisheries was workable. Unrestricted access to fisheries is no longer feasible in the face of accelerating growth in world populations and the subsequent demand for seafood products. Many governments have taken action to control and, in some instances, reduce the numbers of fishing vessels operating in particular fisheries. Some fisheries have been closed, and in others, the length of the fishing season has been curtailed in order to reverse stock declines.

The idea of farming the sea is not new. For hundreds of years the culture of mollusks, crustaceans, fish and seaweeds has been practiced in southeast Asia. Today, southeast Asia produces about two-thirds of the world's aquaculture output, and in some countries in this region, up to 60 percent of the dietary protein comes from 'farmed' organisms. In 1987 world production of marine products through aquaculture totaled only around 3 million tons, or 4 percent of world marine production, by 1993 this had risen to around 6 percent and mariculture farms are becoming an increasingly common sight around the world's coastlines.

Farmed species

The range of organisms cultured varies widely, from the mussel and oyster farms of Europe, through the clam farms of North America, to the seaweed farms of Asia. Shellfish mariculture, such as that of mussels in Europe, can yield harvests of up to 2.5 acres (100 tons of organic food per hectare) per year – considerably higher than that found on land.

In the Philippines, extensive milkfish (*Chanos chanos*) farming was based on fry collected in shallow waters by fishermen using fine nets, who then sell the fry to brackish water farms where they are maintained in shallow nursery ponds before being transferred to larger, deeper ponds. The fish are harvested at all growth stages. The tiny fry are fried and served as a side dish, and the larger adults are eaten as a main course.

Norway is the world's largest producer of farmed salmon with an estimated yield in 1999 of 400,000 tons. The size of the industry is reflected in the fact that 10 percent of Norway's work force is currently employed in fish farming, using the naturally sheltered enclosures of the deep-water fjords which line Norway's coastline. Salmon farming is also of growing importance in Chile, the Faroe Islands, and New

▼ ***Oysters**, Ostrea* edulis, *and mussels have been cultured in Europe and Asia for centuries. This picture shows the Kashikojima pearl beds, Mie Prefecture, Japan. Today, a wide variety of different oyster species are cultured for food and pearl production in different parts of the world.*

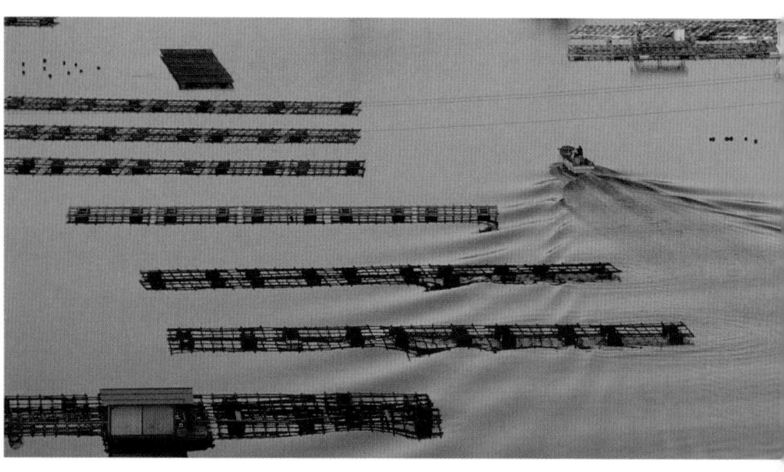

▲ ***In Asia**, seaweed is grown for food and to produce alginates. In this photograph, a man is seen collecting seaweed in the waters around Bali, Indonesia. In recent years, the Balinese have treated seaweed as a crop – planting, tending, and harvesting it in offshore 'fields' such as this.*

Zealand. In 1988 the yellowtail, another important farmed fish species, yielded 200,000 tons worldwide. In Southeast Asia, a wide variety of groupers, sea bass and other tropical species are being farmed, while in the Mediterranean, sea bream and dorade are being farmed in net enclosures.

Some 30 different species of seaweed including sea mustard, kelp and *Euchuema* are grown in East Asia for human consumption, to make alginates used for clarifying, gelling and thickening food products, and as a source of food for animals in culture. Korea and China each produce more than 250,000 tons of seaweed annually. Agar, used in health care as a growth medium for bacteria, derives from seaweed.

Problems of sustainability

Although mariculture is a potential solution to the problems of meeting global demand for seafood products, it can result in severe environmental degradation. Fecal waste and uneaten food result in high rates of bacterial decomposition and deoxygenation of bottom waters. Correct siting of mariculture installations, in areas where water flushing reduces or removes this problem, is now recognized as a necessary prerequisite for sustainable mariculture.

Farming of some species still relies on the unsustainable harvest of wild juveniles, which are reared to marketable size. The construction of many farms involves destruction of the habitats on which the wild populations rely. The clearance of mangrove swamps for shrimp 'farms' destroys the nursery habitat of the juvenile prawns on which the commercial trawl catch depends.

Mineral deposits in the marine environment derive from three major sources: those, such as oil and gas, which have formed in sedimentary deposits on the continental shelf area; minerals eroded from rocks on the land; and finally minerals formed in the ocean itself.

Far less obvious than extraction of oil and gas, but more widespread and of major economic importance, is the recovery of sands and gravels for construction purposes on land. The economic value of such sand and gravel greatly exceeds that of any other mineral type, other than oil and gas, mined from offshore areas.

Sand and gravel

Offshore sand and gravel deposits result from weathering processes on land. The deposits are sorted by the action of currents, so materials in any one location tend to be of a relatively uniform size and density. Gravels extracted from the North Sea include a high proportion of flints weathered from chalk and limestone deposits, which ice sheets have moved into the ocean basin. Subsequent current and water action removed the finer sediments, leaving graded materials suitable for use in a variety of construction aggregates.

Athough not strictly mining, beach replenishment schemes involving pumping sand from offshore onto beaches, are designed to contribute to coastal protection by replacing sand which has been moved offshore by the action of waves

► ***Most offshore metals*** *are found as placer deposits. A metal-bearing rock on land is weathered and the debris produced is washed down to the sea by rivers. There it is sorted by the currents, waves and tides, so that the heavy metal particles accumulate to form deposits of mineral sand. These typically take the form of beach deposits, but where the sea level has changed they can be found well out on the continental shelf. The sands are lifted by dredgers.*

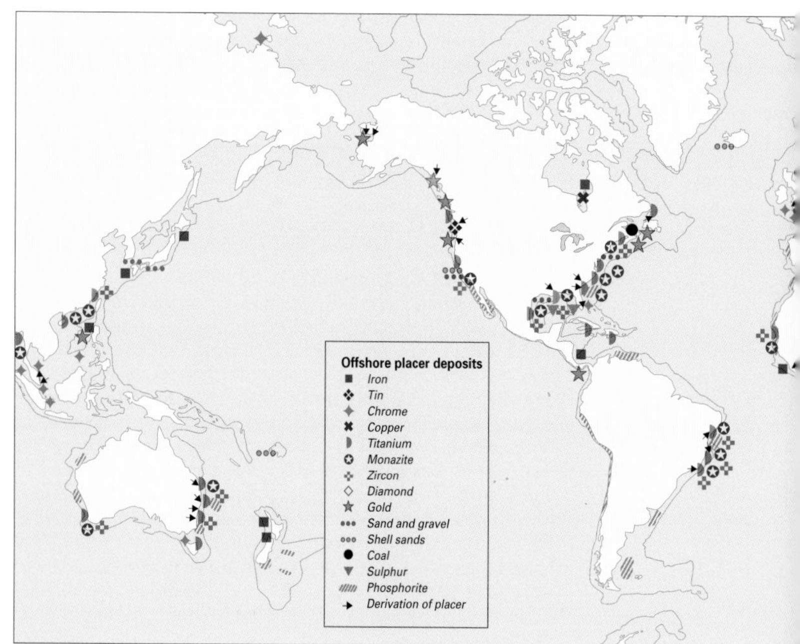

and tidal currents. In small tropical islands, sand for construction is calcareous and is derived from the weathering of coral skeletons. The largest single offshore mining operation involves the extraction of aragonite sands in the Bahamas to be used in the manufacture of high-grade cement. The use of calcareous sands derived from the shells and skeletons of marine animals for the manufacture of cement, is widespread from Iceland to India and relies on the renewable nature of the resource. In Faxa Bay on the west coast of Iceland, for example, the bank of shell sand is constantly replenished by the action of tidal currents and winter storms.

▼ ***With the exception of oil and gas**, most offshore minerals, do not occur in sufficient high ore grades to warrant their economic production.*

Salt and phosphorite

Salt in solution represents about two-thirds of all the minerals in solution in sea water and can be extracted relatively cheaply in countries with a suitably hot, dry climate. In such areas, saltwater is allowed to flood shallow pans where evaporation occurs, leaving deposits of sea salt.

Phosphorite (impure massive apatite) deposits occur in a number of offshore areas, particularly in areas of strong boundary currents and zones of upwelling. Most of the known deposits of phosphorite are geologically fairly old, but some are known to be forming today. Phosphorite is important as a fertilizer and as a chemical agent used in the manufacture of a wide variety of other chemicals. Guano is its best-known form.

Precious metals

For most minerals in the sea (with the exception of salt), the economic costs of extraction exceed the value of the materials. Consequently, most offshore mineral deposits can be considered to be reserves rather than resources, in that prevailing circumstances make them uneconomic. One of the best examples were the diamonds mined from placer deposits off the Namibian coast. Despite the high grade of the deposits, the costs of extraction exceeded the returns, and most operations were suspended in the 1970s. Tin was mined in the 1980s from shallow water in southeast Asia, the ore being dredged from shallow water deposits.

DEEP-SEA MINERALS

▲ ***Manganese nodules*** *were first discovered by the Challenger Expedition (1872–76) and are now known to occur over vast areas of the seabed. Nodules grow very slowly by deposition of minerals around a small object.*

The report of the *Scientific Results of the Exploring Voyage of HMS Challenger* (1872–76) records that "the dredges and trawls yielded immense numbers of more or less circular nodules and botryoidal masses of manganese oxides of large dimensions". This was the first-ever report of deep-sea mineral deposits and led to the well-known term 'manganese nodules'. In reality, the term polymetallic nodule is perhaps more descriptive, since these nodules contain a number of elements in addition to manganese (*see* table).

Until the 1950s, interest in these potential resources was low. The availability of much richer ore deposits on land made exploitation of the polymetallic nodules uneconomic. However, the average grade of copper deposits worked on land fell from around 2.6 percent in 1900 to 0.7 percent by 1965, and the nickel content of ores being mined in New Caledonia fell to about 2.8 percent compared with around 7 percent in 1900. Following World War II, projections of future demand for copper and nickel suggested that lower and lower grade ores would need to be mined. For this reason, interest in the potential mining and extraction of deep-sea manganese nodules increased from the 1960s onward.

Extensive surveys of the world's oceans have shown that nodules ranging from 1–2 inches (2.5–5 centimeters) are widely distributed throughout the deep ocean basins. The

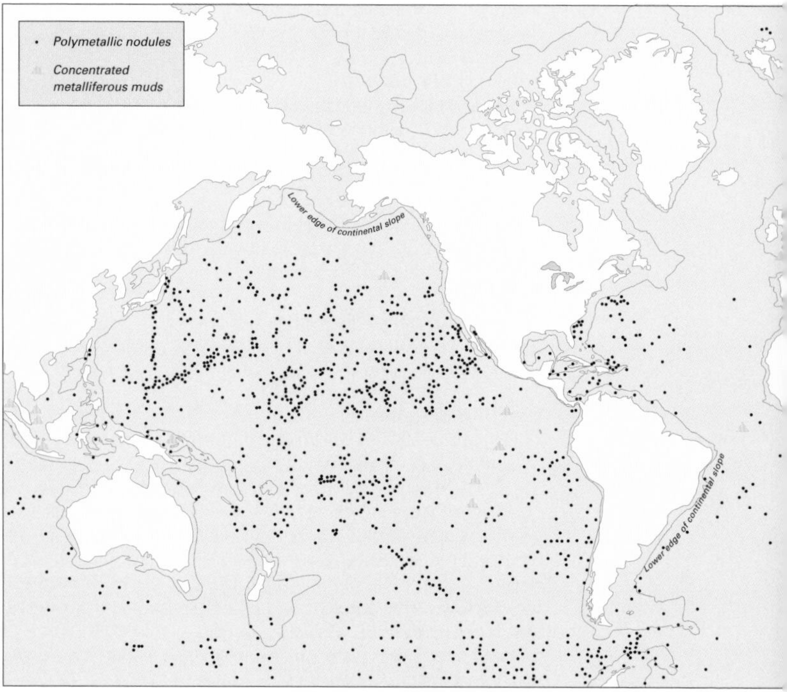

Polymetallic nodules
Concentrated metalliferous muds
Lower edge of continental slope
Lower edge of continental slope

COMPOSITION OF MANGANESE NODULES (AIR-DRIED, PERCENT BY WEIGHT)

Element	Northeast Pacific Ocean	South Pacific Ocean	West Indian Ocean	East Indian Ocean
Manganese	22.33	16.61	13.56	15.83
Iron	9.44	13.92	15.75	11.31
Nickel	1.080	0.433	0.322	0.512
Cobalt	0.192	0.595	0.358	0.153
Copper	0.627	0.185	0.102	0.330
Lead	0.028	0.073	0.061	0.031
Barium	0.381	0.230	0.146	0.155
Molybdenum	0.047	0.041	0.029	0.031
Vanadium	0.041	0.050	0.051	0.040
Chromium	0.0007	0.0007	0.0020	0.0009
Titanium	0.425	1.007	0.820	0.582

▼ ***Most of the world's*** *deep-sea mineral deposits constitute reserves rather than resources, since the costs of their extraction are much too great to permit economic exploitation at the present time.*

richest nodules are generally found in areas away from the input of land-derived sediments and in depths of more than 13,000 feet (4000 meters). The vast majority of ore-grade nodules are now known to occur in the Pacific basin, with the highest density occurring in the Clarion Clipperton Fracture Zone (CCFZ), southeast of the Hawaiian islands in the central eastern Pacific, where there are an estimated 8–25 billion tons of polymetallic nodules.

Mining the seabed

At the present time, the costs of mining manganese nodules are prohibitive. Mining the seabed will involve the collection of the nodules and their transportation to the surface. Unmanned collecting devices could be used for aggregating the nodules, which could be lifted either by suction or by dredging operations. Processing nodules on site or at a submerged processing station would make the mining more efficient, and these possibilities are also being considered.

Vents, brines and metalliferous muds

In the short-term, deposits of metalliferous muds containing varying concentrations of zinc, copper, manganese and lead, are likely to prove more economic. As superheated brines, which may be as hot as 220°F (104°C), are released into the colder water of the deep-ocean floor, precipitation of the dissolved metals takes place and these are deposited as metal-rich mud.

Where water movement is restricted, as in the case of deep pits in the axial valley of the Red Sea, the superheated brine – at temperatures of 140°F (60°C) – fills the pits to a depth of around 650 feet (200 meters). The surface of the ocean floor is covered with metal-rich muds reaching between 6 and 80 feet (2 and 25 meters) in depth. The Atlantis II Deep in the Red Sea is of considerable commercial interest, since the muds in this region contain as much as 40 percent of iron, 3.5 percent of manganese, 2 percent of zinc and 0.95 percent of copper.

OIL AND GAS

The world's continental shelf areas are, in relative terms, a more promising area for oil and gas exploitation than the land. Although the continental shelves occupy only about 10 million square miles (26 million square kilometers), approximately 6 million square miles (15 million square kilometers), or about two-thirds, are composed of the sedimentary basins where oil and gas reserves are normally found. Although the total area of the land is much greater, around 60 million square miles (150 million square kilometers), only about one-third of that area is likely to yield oil and gas.

Oil and gas are formed by the decomposition of organic material in sedimentary rocks. The geological structures suitable for their accumulation must include permeable rocks, such as sandstones and limestones, through which the oil and gas can freely move as they are formed; and cap rocks, such as shale and salt, which seal the deposits in structures, such as anticlines, preventing their escape to the surface or diffusion through surrounding rocks.

Seismic surveys are generally used to determine the structure of the underlying rocks of the continental shelf and serve as the initial means of identifying likely areas for exploratory drilling. A ship towing a ray of sonic sensors fires a 'shot', usually an explosive charge. The sensors detect the reflected

▼ ***Petroleum*** *is formed from decomposed organic material trapped in sedimentary rocks. Once formed, it migrates – along with the gas formed at the same time – through permeable strata until it is trapped by an impermeable layer. Here it accumulates to form a reservoir, with the gas and oil floating on the surface. Permeable rock strata such as red sandstone may be capped by impermeable rocks such as anhydrite or shale, serving as traps for oil and gas.*

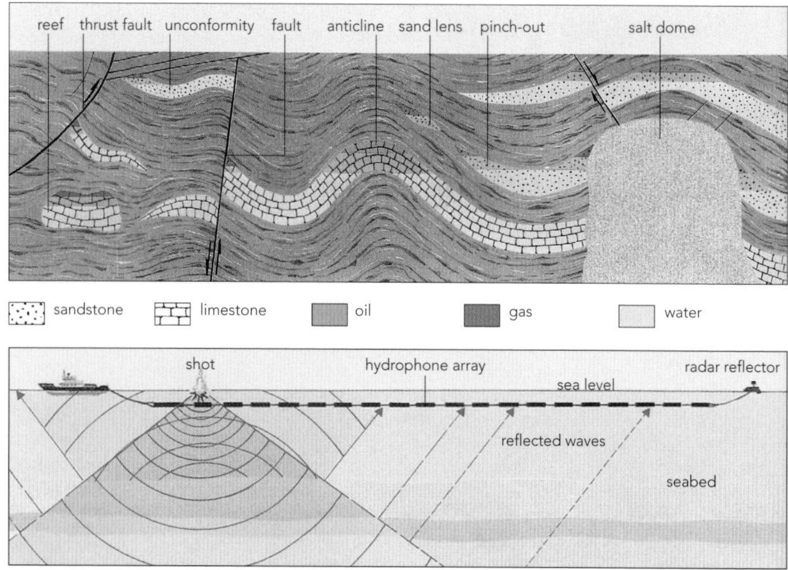

▲ ***Geological structures*** *which might trap oil and gas can be detected by seismic soundings, which provide a picture of this strata. When a seismic shot is fired, the reflected sounds are picked up by sensors such as a hydrophone array. Different rocks and structures reflect the sounds differently, providing a seismic profile.*

waves from the seabed and underlying strata and generate a 'sonic picture' of the geological formations. Following the detection of a likely geological structure, exploratory drilling may proceed. However, a very high proportion of the structures investigated prove to contain either only water or gas and oil reserves that are uneconomic to extract.

Offshore drilling

The first oil to be extracted offshore was from the Summerland field, south of Santa Barbara, western California, in 1896. At this time, wells were drilled from piers that extended up to 800 feet (250 meters) from the shore. During the 1920s and 1930s, wells were drilled in the Baku region of Russia from trestles running out into the Caspian Sea. However, it was the discovery of the large Bolivar oil field in Lake Maracaibo, Venezuela, which really spurred interest in offshore oil reserves. By the 1940s, the first specifically designed offshore steel drilling structure was installed in 23 feet (7 meters) of water in the Gulf of Mexico.

Since that time, the design of offshore oil rigs and production platforms has advanced considerably. The height of these structures has increased as the search for oil has moved further away from the land. By 1975, 77 wells had been drilled in various places around the world in waters of about 650 feet (200 meters) deep.

The growth in offshore oil exploitation is well illustrated in the North Sea where by 1989, 149 platforms were operating, of which 92 were British, 36 Dutch and the remainder Norwegian, Danish and German. By 1990, some 5000 miles (8000 kilometers) of pipeline had been laid, and investment in the North Sea oil industry exceeded US$75 billion. Presently 50,000 people earn a direct living from this industry, which produces more than 150 million tons annually. To date, some 20 percent of the reserves have been exploited, and additional reserves are now known to be present in deeper water off the Faroe Islands.

Initially, offshore oil exploration and extraction relied on transferring land-based technology to the ocean environment. However, sea-based rigs were limited in that they not only had to reach the necessary height above the well head for the successful operation of the drilling equipment, but they also required legs and support structures that were long enough to reach the seabed. One of the tallest platforms in operation today is found near Santa Barbara, California. It stands 1165 feet (355 meters) from the seabed to the top of the derrick – a mere 100 feet (30 meters) or so shorter than the Empire State Building in New York.

From about 1953 onward, rigs were designed that could operate in around 300 feet (90 meters) of water through legs lowered into the seabed on which the drilling platform was jacked up above the sea surface. During the 1960s, the semisubmersible rig was designed. This rig was fit with pontoons that provided buoyancy to a partially submerged structure anchored to the seabed.

Exploration rigs

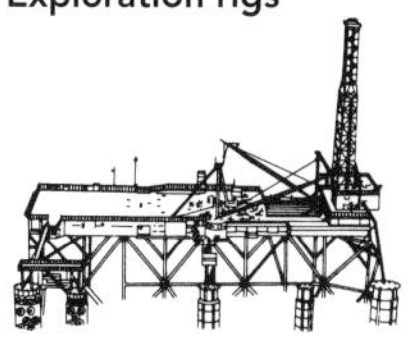

▲ ***Template rig*** *shallow/medium water*

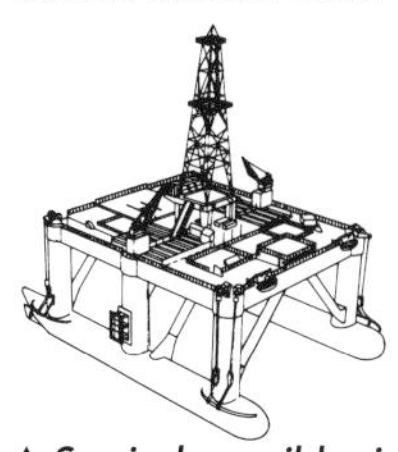

▲ ***Semisubmersible rig*** *deep-water operation*

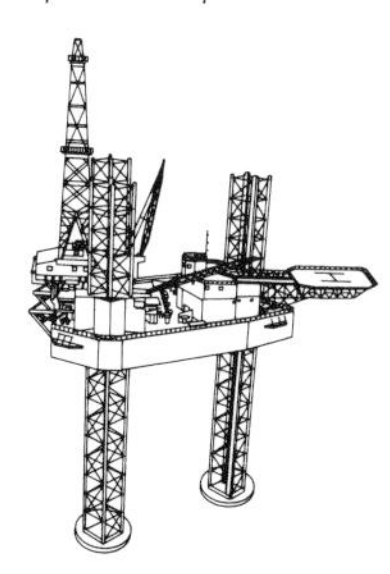

▲ ***Jackup rig*** *medium-water operation*

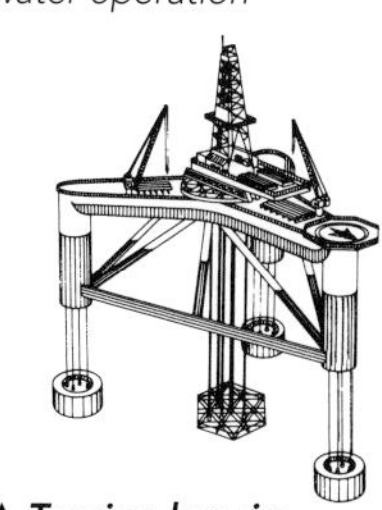

▲ ***Tension-leg rig*** *deep-water operation*

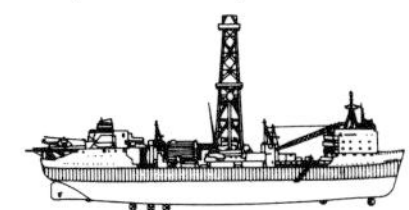

▲ ***Drilling ship*** *general survey*

Mobile rigs

Although drill ships have been used for some time, they were usually anchored to the seabed in order to maintain position over the well head. More recently, however, the development of dynamically positioned drilling rigs has dispensed with the need for such ships to be anchored to the seabed during exploratory drilling operations. The drilling ships maintain their position over the well head by continuous adjustments of specialized propellers, which respond to sonar signals from sources located at known positions around the well head.

The first of these ships was put into service in 1971, and by 1976, the fleet of mobile drilling rigs had increased to 350 worldwide. By the late 1970s, drilling had been carried out in 3300 feet (1000 meters) of water off Thailand, and during the 1980s, test wells were drilled in the Maldives and Philippines at depths in excess of 6500 feet (2000 meters).

▼ ***British Petroleum's*** Bruce *oil drilling platform, located in the northern North Sea, produces 10 percent of Britain's natural gas. The number of offshore platforms for oil and gas production has increased dramatically worldwide over the last few decades, as the demand for petroleum products continues to increase.*

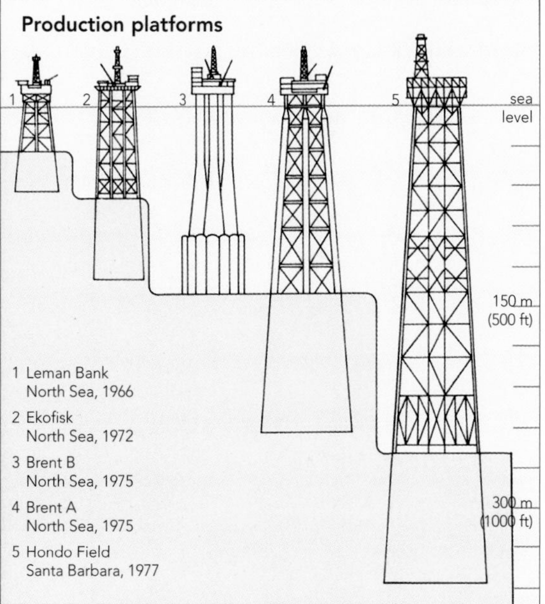

◀ ***The height of oil*** *production platforms has steadily increased as the search for oil and gas extends into deeper and deeper waters. The illustration on the left shows the increase in height of platforms above the seabed from 1966 to 1977.*

The rapid increase in offshore oil exploration and exploitation worldwide reflects the dominance of fossil fuels to the economies of the industrialized nations, and in turn, the increase in demand for this source of energy. In 1960, for example, 90 percent of all offshore oil drilling took place in the United States. By the 1970s, however, the US contribution had dropped to approximately a quarter of world offshore production, a proportion which has further declined as increasing numbers of countries around the world discover and exploit their reserves.

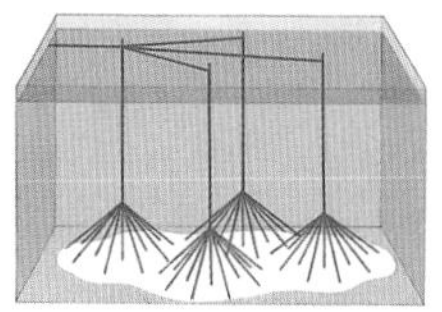

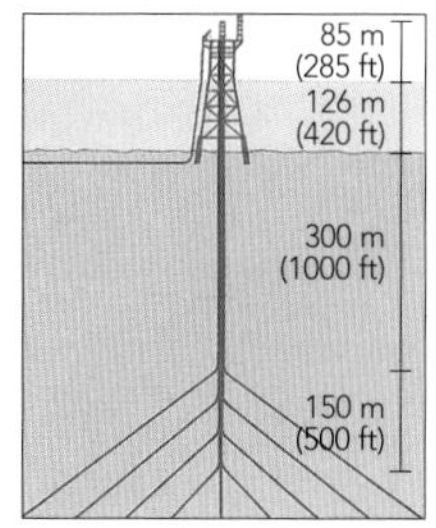

▲ ***The Forties field*** *in the North Sea is worked by four production platforms, each of which can drill 27 wells to 11,500 ft (3500 m). Each well is drilled out to an angle.*

National rights to offshore areas

The value of offshore oil and gas reserves has driven international negotiations aimed at defining coastal states' rights over neighboring marine areas. Following the development of technology during World War II, and as a result of increasing recognition of the potential of oil and gas reserves in offshore areas, the United States issued a unilateral proclamation, declaring exclusive right to exploit its continental shelf. Early negotiations left open the seaward limits to the areas over which a state could claim exclusive rights, and exploitation was limited solely by the available technology. In 1969, the pronouncement of the International Court of Justice in the North Sea Case, recognized that a coastal state has a right to exploit "the natural prolongation of its land mass under the sea". As a consequence, many states lay claim to the whole of their continental margins.

POWER FROM THE SEA

The idea of harnessing the power of the oceans for the production of energy has existed for some time. Yet the efficiency of most systems remains inadequate, making the power they produce more expensive than burning fossil fuels.

Ocean power, in all its many forms, is essentially a renewable energy source. The power derives either from the energy of the Sun, which is stored as heat in the oceans, or it is transferred from the atmosphere to the sea in the form of waves. Tidal energy results from the force of gravity from the Sun and Moon acting on the water of the ocean basins. There are many turbine sites for offshore wind power throughout Europe, and more are currently being developed off the coasts of Britain and Ireland. Offshore wind speeds are usually higher than coastal wind speeds, and each turbine typically generates enough electricity annually to power 1500 households while displacing around 35,000 tons of carbon dioxide.

Wave and tidal power

Wave-power systems are based on the fact that energy is continuously transferred to the ocean from the atmosphere by wind. Wave-power systems also provide coastal protection. In 2000, the world's first commercial wave power station was built on the island of Islay, in Scotland. Known as *Limpet 500* (Land Installed Marine Powered Energy Transformer), the generator collects wave energy in a partially submerged shell. A column of air, contained above the water level, is alternately compressed and decompressed by the waves to generate an alternating stream of air which can be used to produce electricity.

The first fully operational system harnessing tidal power has been working in the Rance estuary, northern France, for several decades. Tidal systems such as at Rance generate electricity

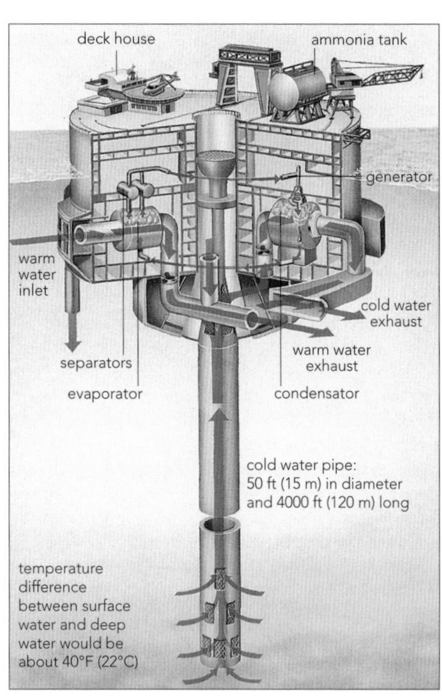

▶ ***Ocean Thermal Energy*** *Conversion (OTEC) is designed to utilize the temperature difference between the warm surface and cold deep waters to drive a closed evaporation condensation cycle based on ammonia or another liquid with comparable thermal properties. The cycle can be used to drive turbines linked to banks of generators.*

through turbines located in a dam spanning the estuary. The turbines are driven by the ebb and flow of water with the tides.

Ocean Thermal Energy Conversion

Some experimental systems, such as Ocean Thermal Energy Conversion (OTEC), are being developed. OTEC depends on the temperature differential between warm, surface water and the considerably cooler water of the deep ocean. The difference may be as much as 36°F (20°C). The cold, deep water is used to condense ammonia or similar liquid chemicals, which are then passed through evaporators warmed by the surface waters. The cycling of the gas through the condensation cycle is used to drive turbines that generate electricity. Although the efficiency of such systems is relatively low, alternative uses for this thermal grade include the Claude Condensator, which can be used for generating freshwater supplies in coastal areas of arid countries.

▲ ***The tidal power*** *station spanning the Rance estuary, northern France, was the first system to harness the power of the ocean.*

▼ ***One design*** *for harnessing wave energy involves huge rocking devices which activate nonreturn valves, forcing water through small-bore pipes to drive turbines.*

Currents and waves

While it would be inefficient to return to the days of sailing ships, ocean currents and waves could be harnessed for maritime transportation. The deep-water outflow from the Mediterranean, for example, is used by submariners to navigate the straits of Gibraltar with engines turned off. It has been calculated that the wave power, which causes a ship to rise and fall at sea, is considerably more than that required to propel it. If this energy could be harnessed, then long-distance transportation costs could be reduced. In addition, small, man-made currents could be used to influence the direction of sediment transportation by tidal currents. More efficient use of such developments would take the place of dredging and costly beach replenishment schemes, both of which rely on fossil-fuel energy.

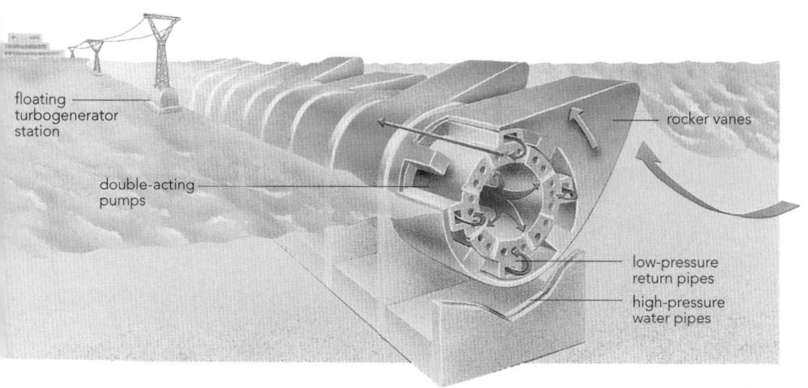

At the time of the United Nations Conference on the Human Environment held in Stockholm in 1972, the greatest threat to the marine environment was perceived to be marine pollution. Waste discharged from ships, as well as urban and industrial effluent from land-based sources, were all seen as contributing to a significant reduction in the health and quality of the marine environment.

Although marine pollution is still considered a major cause of concern, it is now recognized that other human activities, including large-scale commercial fishing and the extensive modification of coastal environments, may be having just as great an environmental impact on the quality of the seas and oceans.

In 1990, a report published by the Joint Group of Experts on the Scientific Aspects of Marine Pollution, which advises the United Nations' agencies, concluded that, "*In 1989 man's fingerprint is found everywhere in the ocean. Chemical contamination and litter can be observed from the poles to the tropics and from beaches to abyssal depths. But conditions in the marine environment vary widely.*" This group further noted in its report in 2001 that, "*population pressure, consumption patterns and increasing demand for space and resources... undermine the sustainable use of oceans and coastal areas and their resources*".

By the time of the United Nations Conference on Environment and Development, held in Rio de Janeiro in 1992, land-based pollution was considered to be the major source of pollution in the marine environment. Of even greater impor-

▶ ***The areas of*** *greatest water pollution in the world. The numbers on the map correspond to major oil tanker disasters, which are listed in the table below.*

▼ ***Satellite image*** *of the northwest coast of Spain, November 21, 2002. Several dark-colored fingers of oil are visible reaching through white waves toward the cliffs along the shoreline. The area pictured here is just north of Cape Finisterre. The oil tanker, containing 77,000 tons of oil, sank some 130 miles (210 km) due west of Cape Finisterre.*

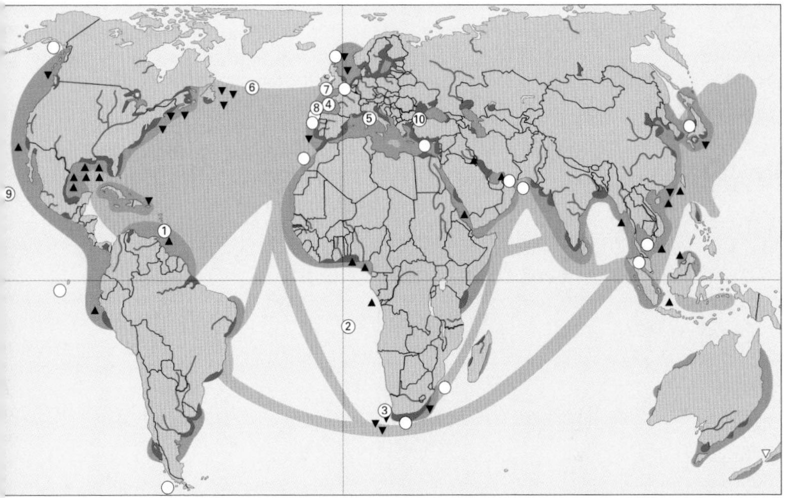

tance, however, was the overriding need to develop more rational management of human uses of the coastal zone and inshore resources. Agenda 21, which was approved by the governments attending the Rio Conference, called for all coastal states to develop integrated coastal zone management plans by the year 2000.

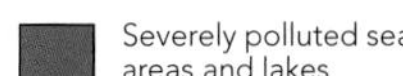
Severely polluted sea areas and lakes

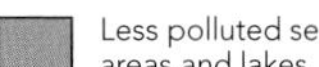
Less polluted sea areas and lakes

Areas of frequent oil pollution by shipping

⑨ ○ Major oil tanker spills

▲ Major oil rig blow-outs

▼ Offshore dumpsites for industrial and municipal waste

— Severely polluted rivers and estuaries

Chemical and oil pollution

In the recent past, heavy metals such as mercury, cadmium and lead were considered among the most pervasive pollutants. It is now recognized, however, that a number of marine organisms naturally concentrate these elements, and hence high concentrations may not necessarily reflect man-made pollution. Nevertheless, pollution by such elements remains a concern in areas of high-industrial discharge, and the well-documented cases of Minamata disease and Itai Itai, caused by eating shellfish contaminated with mercury and cadmium respectively, dramatically demonstrate the threats to public health from discharge of land-based pollutants to the marine environment. Antifouling agents such as organo-tin compounds are known to have major effects on the reproductive biology of shellfish, and some countries have banned their use. Chlorinated hydrocarbon pesticides may also be causing problems along tropical coastlines, although the concentrations of these materials have declined in the developed world after control and restriction of their use.

POLLUTION FROM OIL TANKERS

Tanker name	Tons spilled	Year
1 *Atlantic Express*	287,000	1979
2 *ABT Summer*	260,000	1991
3 *Castillo de Bellver*	252,000	1983
4 *Amoco Cadiz*	223,000	1978
5 *Haven*	144,000	1991
6 *Odyssey*	132,000	1988
7 *Torrey Canyon*	119,000	1967
8 *Urquiola*	100,000	1976
9 *Hawaiian Patriot*	95,000	1977
10 *Independenta*	95,000	1979

▲ ***Oil spills*** *are highly visible forms of marine pollution. Spills such as the* Exxon Valdez *disaster in Alaska had devastating local effects, polluting coastlines and killing valuable fish and marine life. Seabirds such as this jackass penguin, covered in oil after a tanker spill off Western Cape Province, South Africa, are particularly at risk, as the oil causes them to lose their natural bouyancy, and huge numbers can be lost unless the clean-up operation is swift.*

Contaminants enter the sea either through direct discharge or indirectly through rivers and through the atmospheric transportation of particles in aerosols and gases. Around 80 percent of all marine pollution is derived from land-based sources, a further 10 percent results from marine dumping, and the remaining 10 percent from maritime operations, such as ship-based discharge of sewage and ballast water.

Maritime activities have much less impact now than in the past, mainly due to the introduction of international conventions limiting the discharge of wastes at sea. Nevertheless maritime accidents, such as the *Exxon Valdez* and *Prestige* disasters, may have devastating and highly visible local impacts. Although oil may be a highly visible contaminant of the marine environment, particularly following tanker disasters, it is generally of less concern than many other materials. Floating oil is generally less damaging than oil which comes into direct contact with bottom-dwelling organisms, either in the intertidal or subtidal areas. Damage from such accidents is not usually irreversible, although recovery may be slow.

Most of the materials entering the ocean remain in the continental shelf areas and in semienclosed bays and seas where they may be deposited in sediments and resuspended at a later date during storms or dredging operations. In some semienclosed areas, such as the North Sea, the buildup of contaminants has reached unacceptably high levels, resulting in algal blooms, toxic red tides, and viral deaths of marine mammals.

Plastic and other refuse

The haphazard disposal of plastic material on land and from ships at sea results in the fouling of beaches, and it seriously affects marine wildlife, particularly mammals, diving birds and turtles. These animals may become entrapped or tangled in such materials and drown. A considerable quantity of fishing gear is also lost at sea each year, and nylon nets are reported to 'ghost' fish – catching fish years after the nets have been lost. Litter accumulates on beaches and in shallow water habitats and urban debris generally predominates in waters in the vicinity of large cities close to the ocean, while ship-generated litter is a major contributor to the load on remote strand lines. Even pristine environments located far from man-made sources, such as the Southern Ocean, are not free of marine debris.

Nutrient dangers

Present discharges of sewage, both treated and untreated, not only represent a potential health hazard to bathers and seafood consumers, but also perhaps more importantly are increasing the rate of primary production in coastal waters. Sewage and agricultural run-off are high in nitrogen and phosphorus, which encourage phytoplankton production in coastal waters. These high inputs cause rapid growth, or blooms, of phytoplankton, resulting in unsightly algal scum on tourist beaches. When the algae die and sink to the bottom, the resulting bacterial decomposition uses up available dissolved oxygen causing deoxygenation of bottom waters, which in extreme cases will kill fish. Often the species of algae in the blooms produce toxic substances which may be taken up by shellfish, rendering them unfit for human consumption. Changes in phytoplankton communities from diatoms to flagellates have been attributed to declines in silicate inputs to the ocean and concomitant increases in phosphorus.

The River Po in northern Italy discharges some 100,000 tons of inorganic nitrogen and 6000 tons of inorganic phosphorus per year. Nutrient enrichment of the northern Adriatic has become severe, resulting in algal blooms and the deposition of unsightly algal scum on what were once important and attractive tourist beaches.

In open coastal environments, such discharge may not result in immediate problems but the buildup of nutrients and other contaminants in semienclosed seas, such as the Baltic, Black and Mediterranean Seas, can cause enormous future problems. Nutrient inputs to the Black Sea over the last 30 years from the countries of the Danube basin, threaten the very existence of the ecosystem, with ever increasing areas of anoxia of the bottom waters during the summer months.

▼ ***Refuse litters*** *the banks of this coastal community in Sulu, Philippines. Without more appropriate means of garbage disposal, the sea has become a dumping ground for household waste.*

▲ ***A killer whale****,* Orcinus orca, *breaking the surface of the icy sea. A 2001 census counted only 78 southern resident killer whales. Marine animals pay no heed to the boundaries drawn through their world by politicians. Managing wide-ranging migrant species such as whales requires the consent and cooperation of all governments involved if they are not to become extinct.*

The development of international frameworks for managing the oceans and their resources, is based on the concept of the freedom of the seas for navigation. The idea of free access was also applied to the resources of the oceans, which were considered open for use by anyone possessing the means to exploit them.

Such a perspective was acceptable only when the limits of human use did not result in conflict and overexploitation. The rapidly expanding world population and improved technologies for maritime transportation for exploiting the hidden resources of the seabed and for catching fish, have now resulted in levels of use which threaten the long-term sustainability of the ocean environment.

Managing living resources

The history of whaling and many fisheries demonstrate the problems of open access to living marine resources. When a new fish resource is discovered, fishermen invest in boats and gear, and the numbers of boats increase dramatically until a point is reached when the catch starts to decline. Initially, this may result only in a decline in the average size of fish in the population, but this can soon be followed by decline in catches. If the fishing effort remains at the same level, complete collapse of the fishery can occur, with subsequent economic and social impacts for the fishing communities concerned.

In an attempt to address such problems, the concept of the Exclusive Economic Zone (EEZ) was developed, providing a form of ownership that restricts access to the fish stocks of the EEZ to fishermen of the country concerned or licensed fishermen from other nations. Unfortunately, this does not prevent overcapitalization of the national fishing fleet and overexploitation of the resource.

▶ ***School of cardinal fish****,* Rhabdamia sp.*, swimming over a coral reef in the Andaman Sea, southwest Thailand. Many of the world's coral reefs are under threat from the growing coastal populations in tropical developing countries, which depend on these fragile systems for food. Like rain forests, coral reefs are areas of high species diversity. Indeed, they are the most diverse ecosystem in the oceans.*

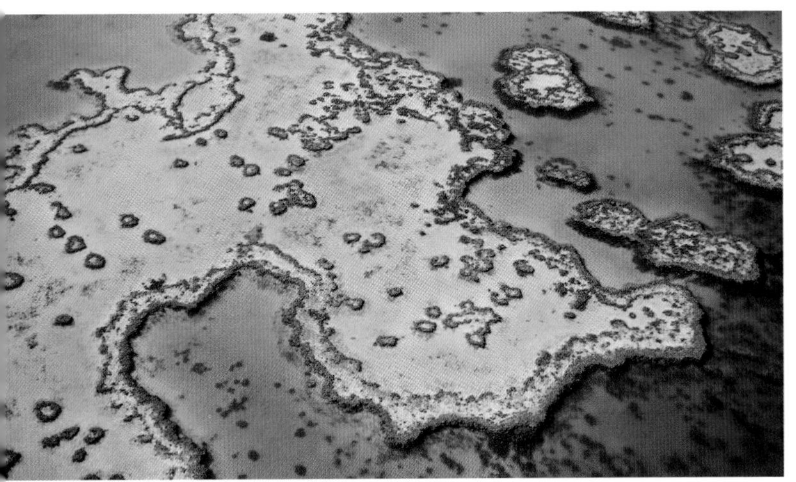

▲ ***The Great Barrier Reef*** *Marine Park off Queensland, northeast Australia, is managed by a single authority, which regulates the use of the area by fishermen and controls the numbers of visitors to offshore islands to reduce the impact of tourism.*

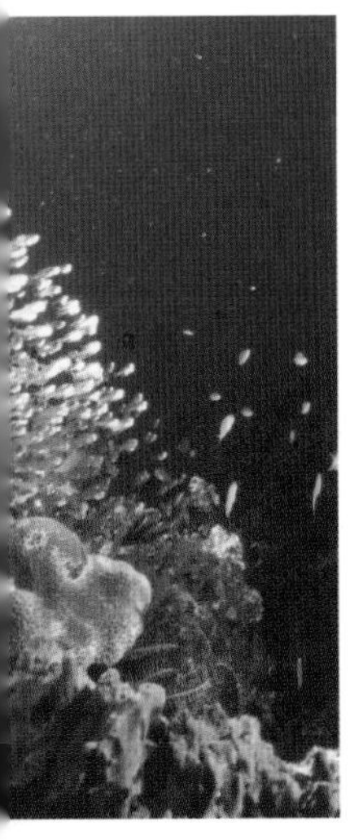

Conflicts of use

Conflicts between different human uses of ocean space are also increasing, and this is particularly apparent in coastal areas. Urban and industrial development conflict with coastal agriculture and fisheries, since the discharge of pollutants affects the health and acceptability of marine foodstuffs, both wild and farmed. Tourist development depends on healthy coral reefs, unpolluted beaches and clean water. However, both mariculture farms and the development of ports and harbors for industry, conflict with such use, since they result in pollution and loss of natural habitat.

Spatial limits and boundaries

A major problem arises when administrative boundaries between towns and villages, or between privately and publicly owned land, separate populations of organisms or functional ocean units. In countries such as Australia, for example, the individual states claim rights to the territorial waters, while the Federal Government claims rights over the 200-mile EEZ. The artificial boundary between territorial seas and EEZ has no biological or physical meaning and may divide stocks or whole ecosystems in a purely arbitrary manner. Similar problems arise when national boundaries divide the same population of fish or when the feeding and breeding areas of a single species are controlled and exploited by two different states. Transboundary stocks that cross on migration, or merely exist in an area divided between two owners, cannot be successfully managed without the joint consultation, consent and agreement of both parties. In recognition of this problem, particularly in relation to pelagic tuna stocks, regional fisheries commissions have been established to provide a forum for negotiation on the fish catches of individual states.

Ocean Atlas

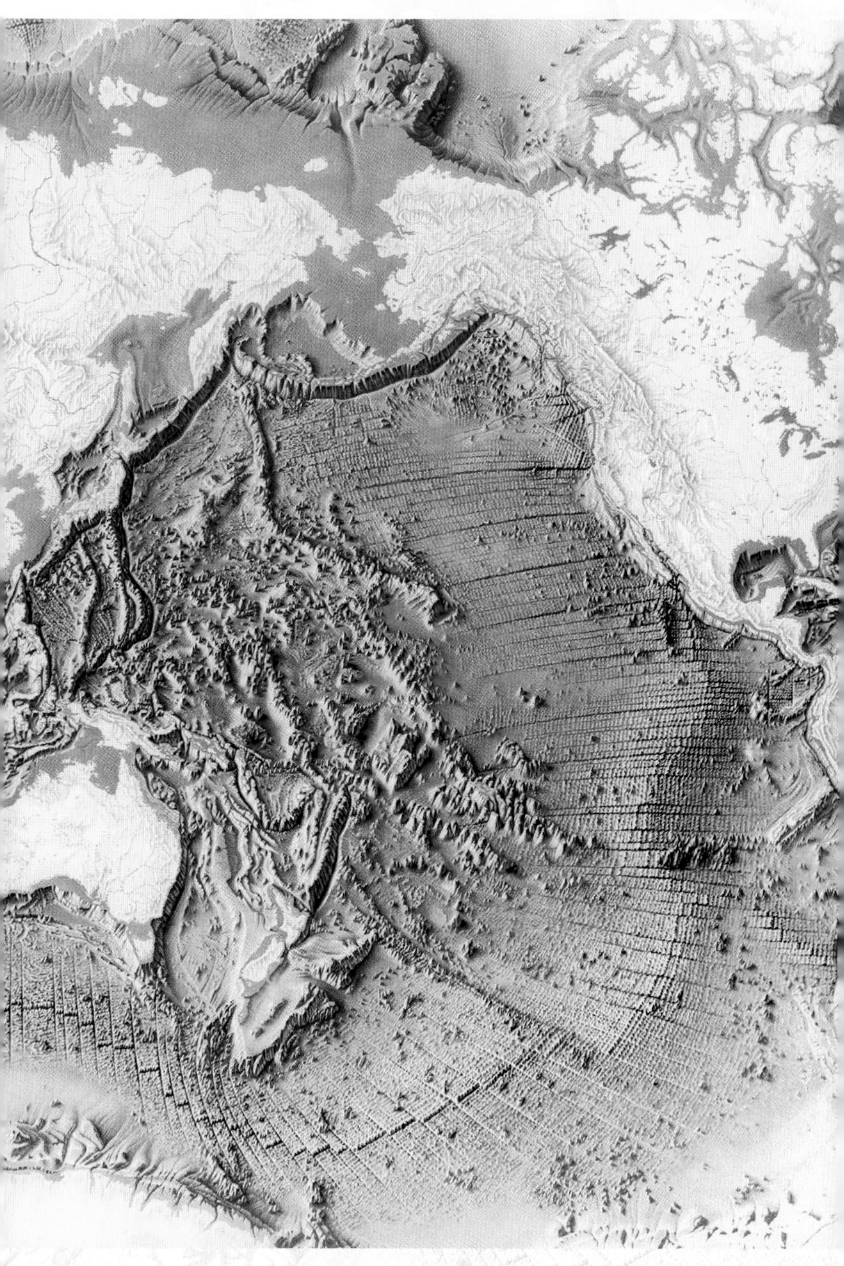

Area: *32,000,000 sq miles (82,000,000 sq km)*
Volume: *77,235,000 cu miles (321,930,000 cu km)*
Average depth: *10,930ft (3330 m)*
Maximum depth: *(South Sandwich Trench) 30,000 ft (9144 m)*

The Atlantic Ocean, although in area some 82 million square kilometers (32 million square miles), is smaller than the Pacific, but it receives freshwater run-off from a far larger area of land; in fact, an area four times as large as that which drains into the Pacific. The saline contents of surface waters in regions of high rainfall and fresh-water run-off from land are generally lower than in other areas. Because the Atlantic receives so much freshwater runoff, the low-salinity waters dominate the surface of much of the Atlantic.

To the north, in the area of the Greenland Sea, the Atlantic is connected with the almost entirely landlocked Arctic Ocean. It is via this channel that approximately 80 percent of water exchange from the Arctic Ocean takes place, and it is here that the cold, high-density Atlantic bottom water forms. Known as the North Atlantic Deep Water, this mass of water sinks and spreads out at a depth of around 5000–13,000 feet (1500–4000 meters), creating a south-flowing, deep-water current that continues south to beyond the equator.

Formation of the Atlantic

The Atlantic first began to form 195–135 million years ago in what is now the central part of the North Atlantic. North America at this time drifted away from the combined landmass of Africa and South America at a rate of about 1 inch (3 centimeters) a year, and about 150 million years ago, the Central Atlantic had opened to roughly 30 percent of its present width.

It was not until the Cretaceous period that the South Atlantic began to form as South America, and Africa began to drift apart. Finally, the extreme North Atlantic was created as the seafloor between Greenland and the Rockall Plateau began spreading about 60 million years ago.

▼ ***The atmospheric** circulation of the Atlantic Ocean closely mirrors the surface ocean currents.*

The Atlantic gyres

Dominating the current pattern of the North Atlantic is the North Atlantic Gyre. An almost circular system of warm, surface-water currents, the North Atlantic Gyre is driven by the atmospheric circulation of the northeast trade winds which blow across the Atlantic between 10° and 30°N; and the Westerlies, found between 40° and 60°N.

In the south, the Antarctic Circumpolar Current carries deep Antarctic bottom waters into the Atlantic, contributing to the strength of the Benguela Current off the west coast of Africa. Unlike

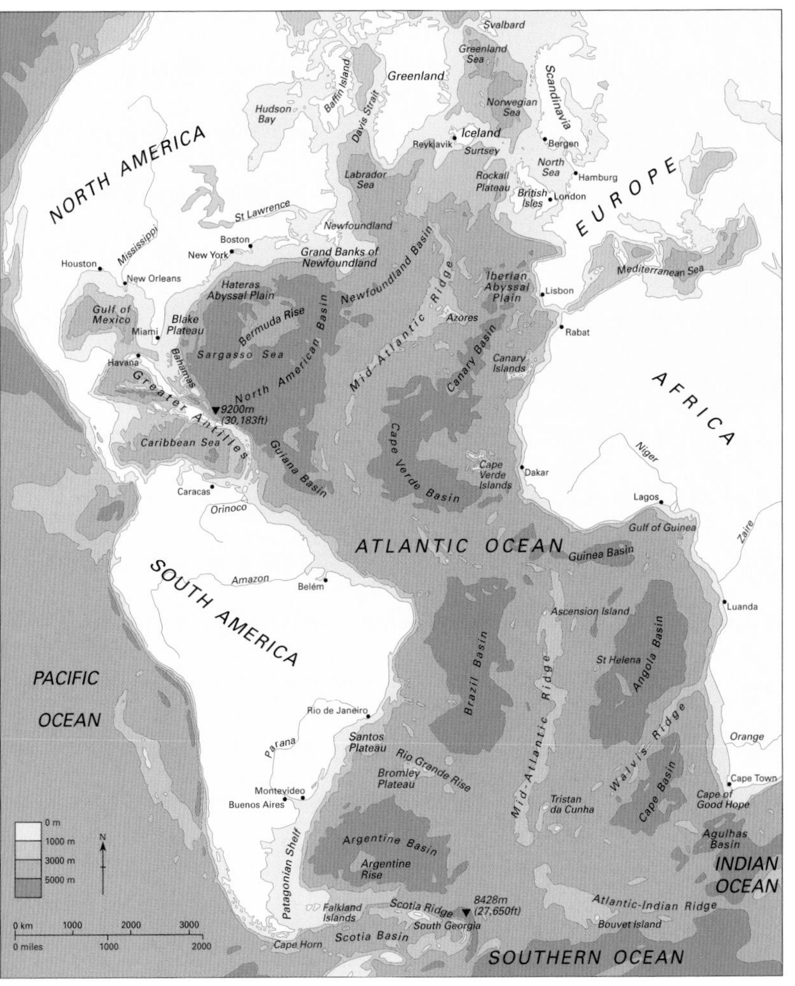

the North Atlantic, the South Atlantic forms a single oceanographic unit. It is also dominated by a warm-water gyre, which again is driven by the atmospheric circulation of the southeast trade winds and the Westerlies.

In contrast, the open ocean waters of the North Atlantic are connected in the west to the semienclosed Caribbean Sea and Gulf of Mexico, while to the northeast the North Sea is connected via a narrow channel, the Skagerrak, to the Baltic. Further south, the Mediterranean Sea receives water from the North Atlantic through the narrow Straits of Gibraltar and is in turn linked to the Black Sea via the Dardanelles and to the Red Sea by the Suez Canal, which opened in 1869.

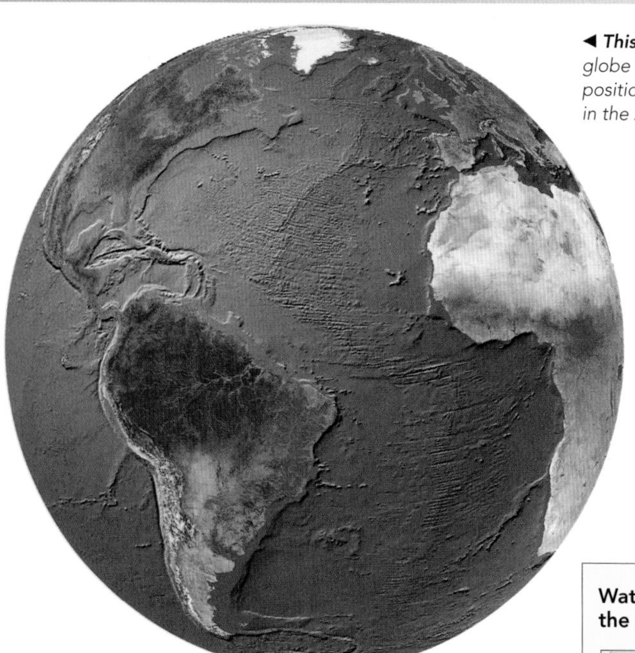

◀ ***This false-color*** *globe indicates the position of the trenches in the Atlantic Ocean.*

The western Atlantic and Sargasso Sea

The surface waters of the western Atlantic are dominated by high salinities and high temperatures as a consequence of the high evaporation and low inputs of freshwater into the Caribbean Sea. These dense, warm waters eventually cool, becoming denser in the northern North Atlantic, sinking to slide south. This combination of density and salinity differences drives the thermohaline circulation pattern of the North Atlantic.

One of the results of the circular motion of the surface currents in a gyre is that toward the center of the gyre the surface waters often have lower motion and indeed may be higher than the surrounding waters balancing the Coriolis force of the surrounding currents. As a consequence, the Sargasso Sea has a mean surface level approximately 1 meter (3 feet) higher than that of the neighboring coasts. Within this area of relatively still water is found a community based on the floating Sargassum weed, related to the brown seaweeds of temperate shores. The Sargassum forms a food source and hiding place for small crustacea, mollusks, sea anemones and fish such as *Histrio*, and a place of attachment for barnacles and algae. This community differs considerably from that of the surrounding waters, since the large size of the algae, compared with the plankton of the open ocean, provides a major food resource for herbivorous animals.

Water profiles of the Atlantic Ocean

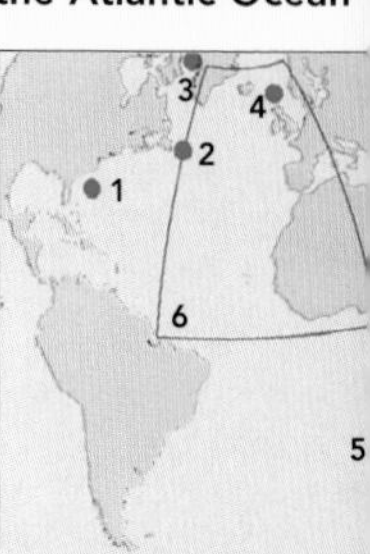

Location of water profile

The warm waters of the Gulf Stream form the northern boundary of the North Atlantic Gyre, below which North Atlantic Deep Water can be seen flowing south to a region beyond the equator. This mass of cold water creates a south-flowing current along the western margin of the ocean.

Not only does the Sargasso Sea form a home for the permanent seaweed community, but this is also the area to which freshwater eels from North America and western Europe migrate on their annual spawning migrations. The larval eels are carried by the currents of the North Atlantic Gyre northward along the Atlantic seaboard of the United States and across to western Europe.

THE ATLANTIC OCEAN BASIN

The Atlantic Ocean floor is dominated by an S-shaped, mid-ocean ridge. This vast, underwater mountain range extends from north of Iceland as far south as Bouver Island on the margin of the Southern Ocean. Known as the Mid-Atlantic Ridge, it divides the Atlantic Ocean Basin into two parallel troughs, which in turn are subdivided by transverse ridges. Some peaks along the ridge emerge as the islands of the Azores, Ascension Island and Tristan da Cunha, but most of the ridge lies 1–2 miles (1.5–3 kilometers) below the surface. Along the center of the ridge runs a deep rift valley, which varies in width between 15 and 30 miles (24 and 48 kilometers)

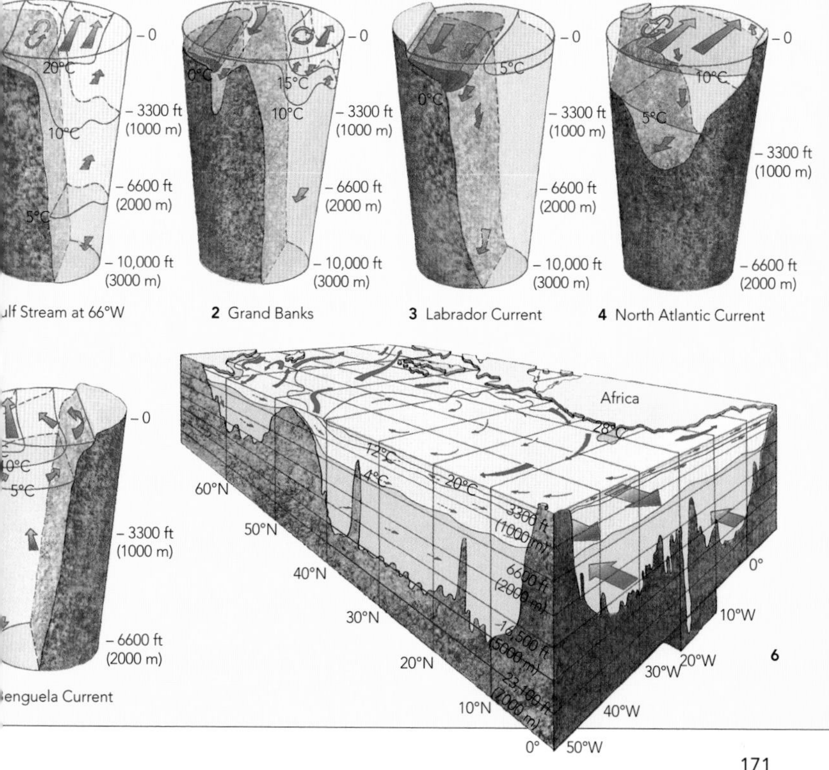

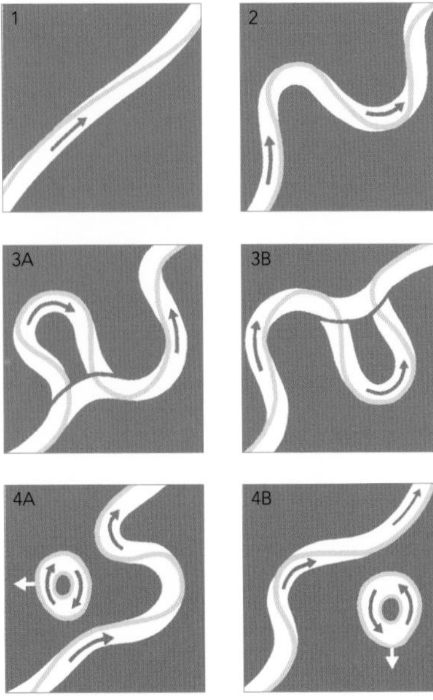

▲ ***Growing instability*** *causes Gulf Stream meanders to increase in amplitude (1–2) until they break away as eddies. If the meanders break away on the northern side of the Gulf Stream (3A), then the eddy has a warm core of Sargasso Sea water and rotates anticyclonically (4A). If, however, the meander breaks away to the south (3B), the eddy rotates cyclonically around a cold core of water (4B). North-side eddies progress westward to rejoin the eastward-flowing Gulf Stream off Cape Hatteras in North Carolina, and become part of the northern boundary of the North Atlantic Gyre. South-side eddies make a long loop to rejoin the Gulf Stream near Florida.*

The valley is displaced by east-west transform faults, which can be seen as narrow ridges and deep clefts, in some places extending more than 1250 miles (2000 kilometers) from the center of the ridge.

The Atlantic Ocean is continuing to expand at a rate of 0.4–0.8 inches (1–2 centimeters) per year, due to the process of seafloor spreading. The volcanic eruption south of Iceland in 1963, which formed the island of Surtsey, was a dramatic example of this process in action. Passing away from the midocean ridge toward the continental rise, the two troughs become progressively deeper, and the lateral ridges and clefts become obscured by increasing depths of sediment brought down from the adjacent land and continental shelves by bottom currents.

Sediment eroded from the land and brought down into the coastal ocean flows across the continental shelf and into the ocean basin via deep canyons cut into the face of the continental rise. At the foot of the continental rise, the sediment may be redistributed by bottom currents to accumulate in specific regions, such as the Argentinian Rise, which lies northeast of the Falkland Islands.

At the edge of the continental slope, land-derived sediments may exceed 3 miles (5 kilometers) in thickness, and off the North American and North African coasts, the sediments date back as far as the Jurassic period (170–160 million years ago). In contrast, the depth of sediment over the younger sea floor is usually less than 0.6 miles (1 kilometer) deep and consists of pelagic ooze and clay.

Currents and circulation

Perhaps the most familiar current of the North Atlantic is the Gulf Stream – a strong, narrow river of warm water which flows north at more than 80 miles (130 kilometers) a day. It flows along the eastern seaboard of North America before leaving the shore around the latitude of the westerly winds (40–60°N) and crossing the Atlantic as the North Atlantic Current, the northern boundary of the warm-water North Atlantic Gyre. Without the influence of the Gulf Stream, the winter climate of western Europe would be much more severe than at present.

A second warm-water gyre lies in the south. The South Atlantic Gyre is formed by the northern portion of the Antarctic Circumpolar Current, which flows along the eastern coast of Africa as the Benguela Current before crossing the Atlantic as the Equatorial Current. This current divides at the coast of Latin America with the northern branch sweeping into the Caribbean Sea as the Guiana Current and the southern branch passing southward along the coast as the weak Brazil Current.

Separating the northern and southern warm-water gyres is the Equatorial Countercurrent, which flows on the surface toward the North African coast. This current overlies the substantial Equatorial Undercurrent, a body of water about 125 miles (200 kilometers) wide, which flows east at about 50 miles (80 kilometers) a day at a depth of 330 feet (100 meters).

At higher latitudes, the westerly winds drive two cold-water gyres. In the north, the subpolar gyre consists of the Irminger Current, the Greenland Current and the North Atlantic Current. In the south a similar cold-water gyre is found in the area of the Weddell Sea.

▼ ***The movement** of the Atlantic currents, and the Gulf Stream in particular, is a vital factor in determining temperatures in the Atlantic region.*

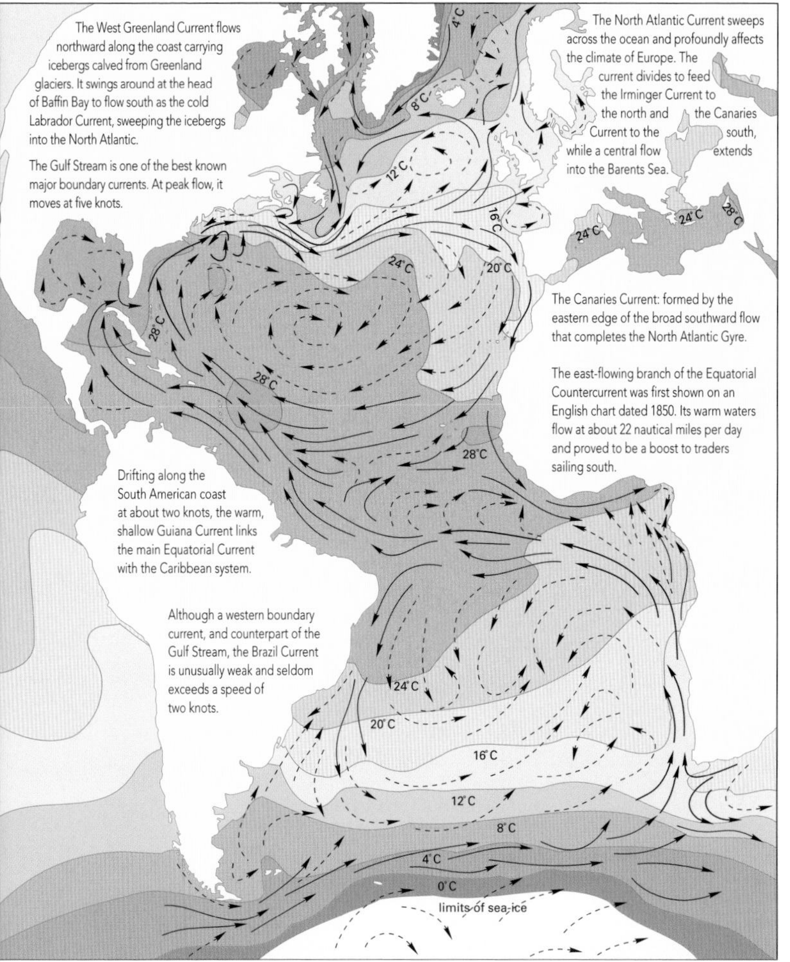

The West Greenland Current flows northward along the coast carrying icebergs calved from Greenland glaciers. It swings around at the head of Baffin Bay to flow south as the cold Labrador Current, sweeping the icebergs into the North Atlantic.
The Gulf Stream is one of the best known major boundary currents. At peak flow, it moves at five knots.
The North Atlantic Current sweeps across the ocean and profoundly affects the climate of Europe. The current divides to feed the Irminger Current to the north and the Canaries Current to the south, while a central flow extends into the Barents Sea.
The Canaries Current: formed by the eastern edge of the broad southward flow that completes the North Atlantic Gyre.
The east-flowing branch of the Equatorial Countercurrent was first shown on an English chart dated 1850. Its warm waters flow at about 22 nautical miles per day and proved to be a boost to traders sailing south.
Drifting along the South American coast at about two knots, the warm, shallow Guiana Current links the main Equatorial Current with the Caribbean system.
Although a western boundary current, and counterpart of the Gulf Stream, the Brazil Current is unusually weak and seldom exceeds a speed of two knots.
4°C
8°C
12°C
16°C
20°C
24°C
28°C
24°C
28°C
24°C
28°C
28°C
24°C
20°C
16°C
12°C
8°C
4°C
0°C
limits of sea-ice

ATLANTIC RESOURCES

▲ ***A test drilling rig*** *explores for oil in the Atlantic Ocean off the coast of Atlantic City, New Jersey, USA.*

Most of the mineral deposits of the Atlantic, with the exception of oil and gas, are of insufficient economic value to merit commercial exploitation. However, some of the finest diamonds in the world are found on the marine terraces of Oranjemund, Namibia, off the southwest coast of Africa, and gem-quality diamonds are recovered from raised beaches and placer deposits lying below the tideline.

Production of those minerals, which are currently being exploited, is largely concentrated in shallow water on the continental shelf close inshore. Petroleum and gas extraction is centered on the wider Caribbean region, the North Sea, and along stretches of the West African coast.

Apart from oil and gas, the value of sand and gravel is greater than that of all other minerals combined, and most of this extraction occurs off the coast of northwest Europe, where around 10 million tons of gravel are dredged annually by operations in the North Sea. Dredging aragonite sands from the Great Bahamas Bank is also of commercial importance.

Depleted living resources

The majority of the North Atlantic fish catch comes from the continental shelf areas bordering northwestern Europe, eastern Canada and the United States. The North Atlantic is the world's most heavily fished ocean area, and numerous fish stocks are suffering from overfishing. The North Atlantic salmon fishery has declined substantially in recent decades. Stocks of salmon and steelhead in the Columbia River basin, for example, have declined by 80 percent from historic levels, while Californian salmon stocks are down by 65 percent. Although salmon spawn in rivers on both sides of the Atlantic, much of their adult life is spent off Greenland, where commercial fishing has resulted in depletion of the stocks. Much of the caught wild salmon has now been replaced by farmed salmon, particularly from Scandinavia.

The most important fisheries by weight of catch in the North Atlantic are the pelagic species, dominated by sardine and anchovy. Although not as large as that of the Pacific, the Atlantic tuna fishery is also significant. Demersal fish, particularly cod, flounder and plaice in the north, and hake from southern Europe, northern Africa and America, are also heavily fished, leading to recent disputes between Canada and the European Union regarding cod fishing on the Grand Banks. Atlantic finfish catch in Canada, for example, has declined

dramatically over the last 30 years from a peak of around 2.5 million tons in the early 1970s to less than half a million tons in the early 1990s. In the early 1990s, Canada imposed a two-year moratorium on northern cod which has since been extended indefinitely, while in 1993 the US imposed stricter limits and shorter seasons.

Crustacean fisheries for lobster, prawn and shrimps are widespread, with 4000 lobster boats operating in the Gulf of Maine. Most of the lobsters come from the northeast United States, the Caribbean, northern Brazil and along the South African coast. Penaeid shrimps are fished off the West African coast and in the Caribbean, while crabs form an important resource off the eastern seaboard of the United States and to a lesser extent in the North Sea.

The rich fisheries of the West African coast are the consequence of high primary production resulting from the upwelling of cold, nutrient-rich waters off Senegal and the Congo. In addition, the high nutrient inputs from the Congo drainage basin also contribute to the high productivity of this area.

*▼ **International cod** evels have declined dramatically in the last ew decades. In 1992, he cod population in Newfoundland, Canada, was found to be at one percent of its size in he 1960s. International aws now regulate catch sizes. In 2001 cod accounted for 42% of celand's total seafood export revenue.*

Farming the sea

Mariculture is widespread in the North Atlantic, with salmon and trout being farmed in Canada, Scandinavia and western Scotland, and tuna in Nova Scotia. European mariculture of marine fish reached 34,235 metric tonnes in 1995, compared with only 3,310 tonnes in North America and a world total of 472,243 tonnes. Production of Pacific oysters in France exceeded 125,000 metric tonnes in 1985, while in the same year, Spain produced more than 240,000 metric tonnes of blue mussels. Mariculture of shellfish in Europe reached 623,426 metric tonnes in 1995, compared with only 162,664 tonnes in North America and a total world production of nearly 6 million metric tonnes. Mariculture of oysters, clams, mussels and various fish expanded in the United States throughout the 1990s; however, degradation of water quality may limit further growth of this industry along the coastlines of the industrialized countries.

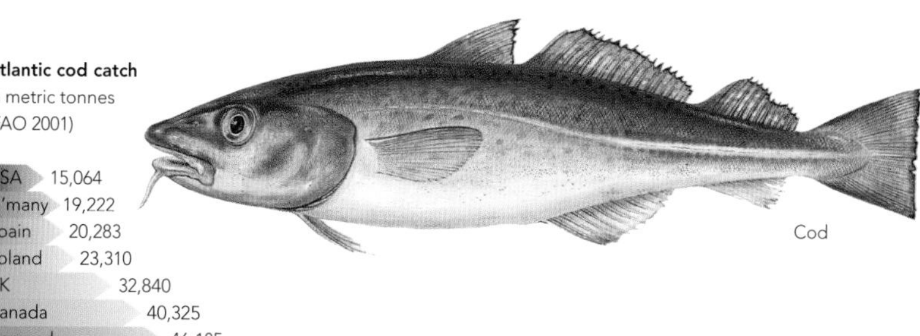

Cod

The Gulf of Mexico
Area: *598,000 sq miles (1,543,000 sq km)*
Volume: *560,000 cu miles (2,322,000 cu km)*
Average depth: *4960 ft (1512 m)*
Maximum depth: *13,218 ft (4029 m)*
The Carribean Sea area: *1,020,000 sq miles (2,640,000 sq km)*
Max. depth: *(Cayman Trench) 25,216 ft (7686 m)*

The wider Caribbean region is located in the southwestern margin of the North Atlantic, but separated from it by the Bahamas and the Greater and Lesser Antilles. The region, which includes the Caribbean Sea and the Gulf of Mexico, is a relatively young ocean area, which only assumed its present form some three million years ago following the closure of the Isthmus of Panama.

The sliding sea floor

The ocean floor is divided into four major basins: the Gulf of Mexico to the north, the Yucatan Basin in the center and the Colombian and Venezuelan Basins to the south which, together with the Grenada Trough to the east, form the Caribbean Sea. The floor of the Caribbean Sea is believed to be a fragment of the Pacific Ocean crust cut off at the time that South America joined Central America.

To the east, the Atlantic crust slides beneath the Lesser Antilles, resulting in volcanic eruptions such as that at Mount Pelee, which killed 30,000 inhabitants on St. Pierre, Martinique

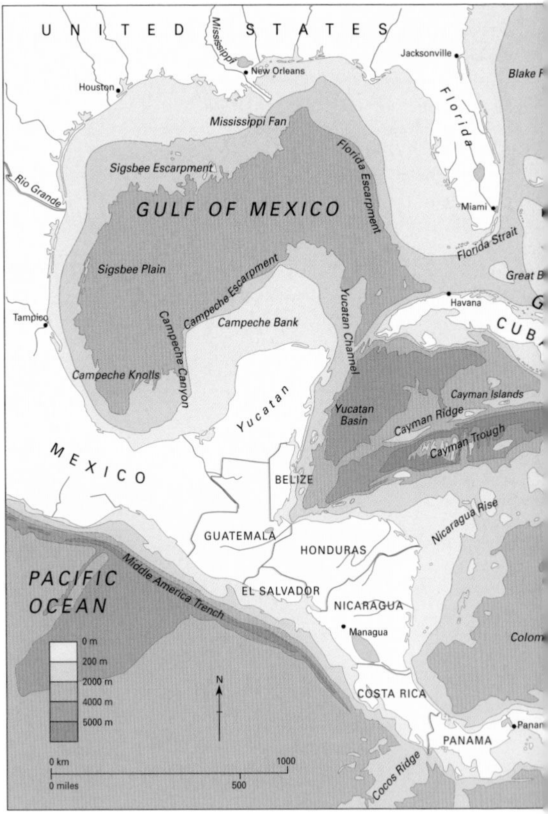

► **Map** *showing the major basins that make up the Carribean Sea.*

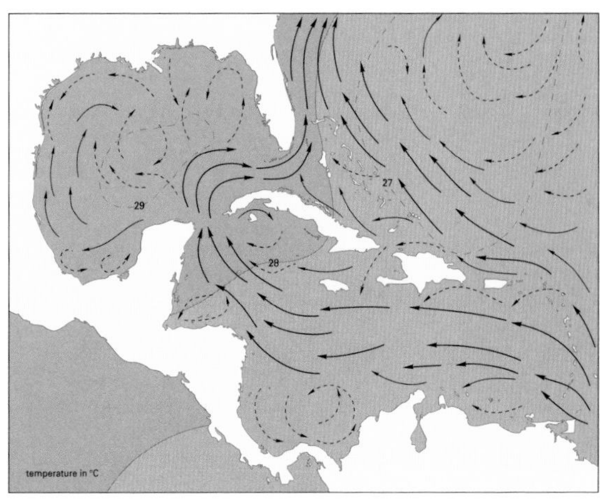

▶ ***The circulation*** *of the Caribbean's surface waters is dominated by the warm Gulf Stream. The Gulf Stream flows out of the Gulf of Mexico and then follows the northeastern coast of America, before passing across the Atlantic and on to western Europe.*

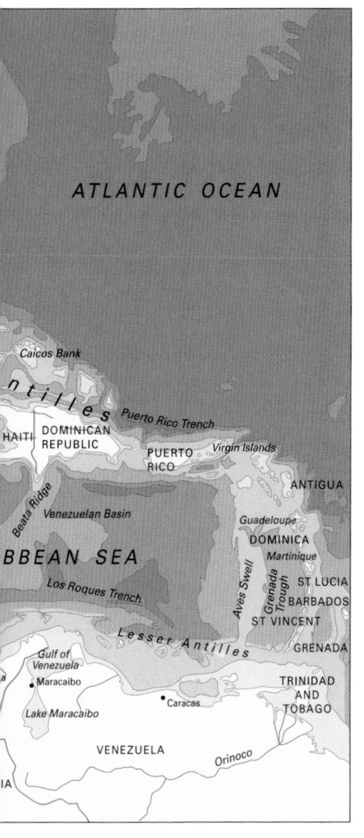

on May 8, 1902. To the west, the Pacific crust slides beneath the landmass of Central America, and violent earthquakes and volcano activity also occur here. Along the northern edge connecting these two zones is a giant fault, along which North America is sliding westward past the Caribbean plate. A movement of around 10 feet (3 meters) along this fault caused the Guatemala City earthquake in 1974. To the south along the north coast of South America, the faults are more complex, although the general direction of movement of South America, like North America, is westward.

Basin divisions

The Aves Swell, which separates the Grenada Trough from the Venezuelan Basin, is composed in part by rocks of continental origin, while the Beata Ridge which separates the Venezuelan and Colombian basins, is in contrast a section of ocean crust which appears to have been uplifted at the time of the separation of the Caribbean plate from the Pacific. The Nicaragua Rise, however, is geologically more complex and of uncertain origin.

Circulation

The surface water of the top 100 meters (30 feet) or so of the Caribbean Sea behaves as an extension of the North Atlantic. The Guiana Current and part of the North Equatorial Current flow past St. Lucia into the Caribbean, and continue westward at a speed of around 20 miles (32 kilometers) per day.

In the western Caribbean Sea, the trade winds cause the surface currents to flow northward away from the South American coast, drawing up colder nutrient-rich water from around 650 feet (200 meters) that supports a rich fishery in the area.

In the Yucatan Basin, surface water flows north through the Yucatan Channel into the basin of the Gulf of Mexico where it is forced to the east toward the Straits of Florida. This turn is sharp, and current meanders may become cut off as warm-water eddies which drift westward across the Gulf of Mexico. The heating of the surface waters as they flow through the Caribbean and Gulf of Mexico regions, contributes substantially to the warming of the Gulf Stream.

The Caribbean's deeper water

The deep basins of the wider Caribbean region are as deep as much of the North Atlantic, up to 16,400 feet (5000 meters). They are, however, separated from the Atlantic by the island arcs and from each other by the ridges. They contain water at roughly the same temperature – 40.73°F (4.85°C) – as the Atlantic Ocean at a depth of 5250 feet (1590 meters), and it seems likely that this Atlantic water spills over the ridges periodically to form the Caribbean Deep Water.

The separation of the Caribbean from the Atlantic is also reflected in the tidal pattern. In the Caribbean, tidal range is smaller than in the Atlantic, and the dominant tide is diurnal, giving a single high and low each day. In the Atlantic, the dominant tide is the semidiurnal, giving two high and two low tides each day.

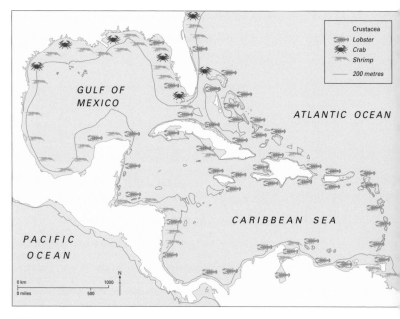

Crustacea
Lobster
Crab
Shrimp
200 metres
GULF OF MEXICO
ATLANTIC OCEAN
CARIBBEAN SEA
PACIFIC OCEAN
0 km 1000
0 miles 500
N

▲ ***The world's largest*** *artificial reef was created in 1996 from a disused sulfur mine off Grand Isle, Louisiana, USA. Artificial reefs encourage fish to breed in the vicinity. In 1963 Garden Island Bay and the Grand Isle mines yielded more than 2.5 million tons of sulfur, making Louisiana the largest sulfur producer in the world.*

CARIBBEAN RESOURCES

The Caribbean Sea and Gulf of Mexico have a long and complex sedimentary history, and this, combined with the restricted water circulation between the basins, has resulted in varied and valuable mineral resources.

Oil and gas

Since the late 1930s, the Gulf of Mexico has become an important center for the extraction of oil and natural gas. Important oil and gas fields lie off the coast of Louisiana and the Mexican states of Veracruz and Campeche, as well as Venezuela and the island state of Trinidad and Tobago. The development of these resources, however, has led to increasing levels of contaminants in the Gulf, with unfortunate consequences for the fisheries of the area. The fisheries that have been worst affected are those that are dependent on the region's extensive mangrove systems.

Sulfur deposits

The north and western margins of the Gulf of Mexico contain large evaporite beds of sedimentary rocks, formed when seawater evaporated during earlier geological periods, leaving salt deposits. These evaporite beds contain domes and plugs, both of which are formed from volcanic activity during which the magma of the volcano solidifies before it reaches the surface. The domes and plugs contain significant amounts of salt, potash and magnesium, while the anhydrite beds comprise reserves of sulfur (in the form of calcium sulfate), which occurs as cap rocks over the salt domes. The world's first offshore sulfur mine began operation in 1960, 7 miles (11 kilometers) off the coast of Louisiana. The sulfur is extracted by the Frasch process in which hot water is pumped down the boreholes under pressure, causing the sulfur to melt. The sulfur is then pumped out in liquid form.

Placer deposits of monazite, zircon and titanium are known to occur around the northern margins of the Gulf of Mexico, while iron and titanium are present in continental shelf deposits off the coasts of Costa Rica and Colombia. Placer deposits of chromite, titanium and gold are also present in the shelf sediments off Cuba and Haiti.

◀ ***The dominant fishery*** *of the Caribbean region is for shrimps. It centers on the Gulf of Mexico. Fishing for both shrimp and blue crab has been carried out in the region's coastal and estuarine waters for many thousands of years.*

Inshore fishing

The Caribbean region is characterized by its small-scale, commercial and artisanal fisheries that reflect the ecological complexity of the region, which contains some 14 percent of the world's coral reefs. Considerable areas of mangrove forest, salt marshes and sea grass beds contribute to the inshore productivity, and as a result of this diversity of habitats, the fisheries are characterized by a high diversity of species in the catch, including groupers, snappers, jacks, flying fish and tuna.

The Arctic Ocean
Area: *4,700,000 sq miles (12,173,000 sq km)*
Average depth: *3250 ft (990 m)*
Maximum depth: *(Pole Abyssal Plain) 15,091 ft (4600 m)*

The Arctic Ocean is the smallest ocean in the world. It is surrounded by Eurasia, Greenland and North America, and is, on average, approximately 3300 feet (1000 meters) deep – although in places it can reach six times this depth. There are four major basins separated by three oceanic ridges of which the largest, the Lomonosov Ridge, extends northwest to southeast for 1100 miles (1750 kilometers) and rises 10,000 feet (3000 meters) above the Pole Abyssal Plain to reach within 3600 feet (1100 meters) of the surface.

Shallow seas

The continental shelf regions of the Arctic are one of the ocean's most unusual features. While the continental shelf of the Canadian and Alaskan coasts is of average width, generally between 30 and 75 miles (50 and 125 kilometers), the continental shelf off the north Asian coast is considerably wider, extending out for more than 1000 miles (1600 kilometers) at its widest, and nowhere is it less than about 300

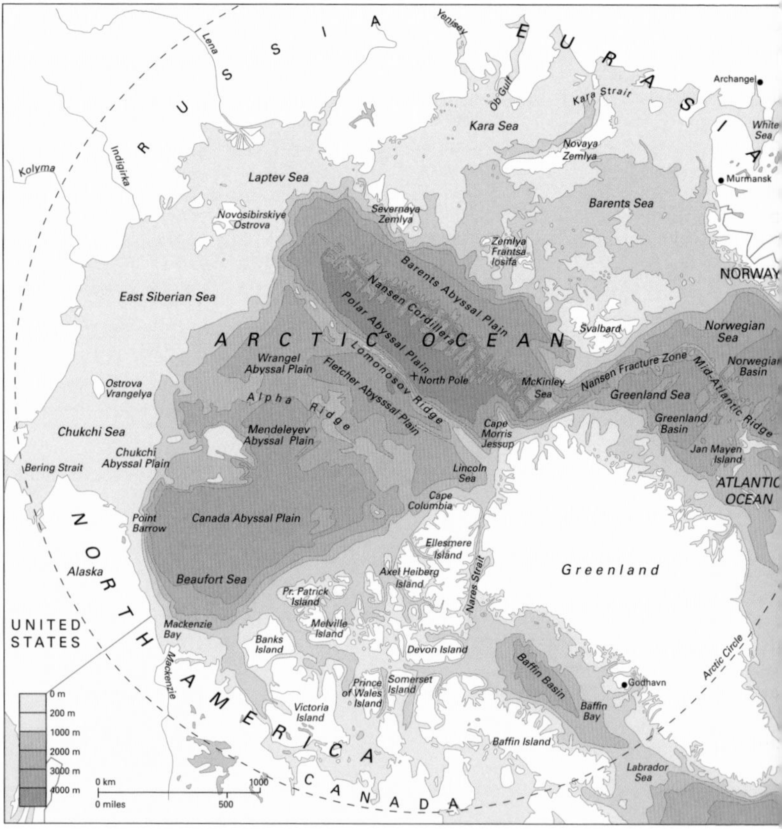

EURASIA
RUSSIA
Lena
Yenisey
Ob Gulf
Kara Strait
Archangel
White Sea
Kara Sea
Novaya Zemlya
Kolyma
Indigirka
Laptev Sea
Murmansk
Barents Sea
Novosibirskiye Ostrova
Severnaya Zemlya
Zemlya Frantsa Iosifa
NORWAY
East Siberian Sea
Barents Abyssal Plain
Nansen Cordillera
Polar Abyssal Plain
Svalbard
Norwegian Sea
ARCTIC OCEAN
Wrangel Abyssal Plain
Lomonosov Ridge
Norwegian Basin
Nansen Fracture Zone
Mid-Atlantic Ridge
North Pole
McKinley Sea
Greenland Sea
Fletcher Abysssal Plain
Ostrova Vrangelya
Alpha Ridge
Greenland Basin
Chukchi Sea
Mendeleyev Abyssal Plain
Cape Morris Jessup
Jan Mayen Island
Chukchi Abyssal Plain
Bering Strait
Lincoln Sea
ATLANTIC OCEAN
Cape Columbia
Point Barrow
Canada Abyssal Plain
Ellesmere Island
Alaska
Greenland
Axel Heiberg Island
Beaufort Sea
Nares Strait
Pr. Patrick Island
UNITED STATES
Mackenzie Bay
Melville Island
Banks Island
Devon Island
Baffin Basin
Arctic Circle
Mackenzie
Prince of Wales Island
Somerset Island
Godhavn
NORTH AMERICA
Victoria Island
Baffin Bay
Baffin Island
Labrador Sea
CANADA
0 m
200 m
1000 m
2000 m
3000 m
4000 m
0 km
1000
0 miles
500

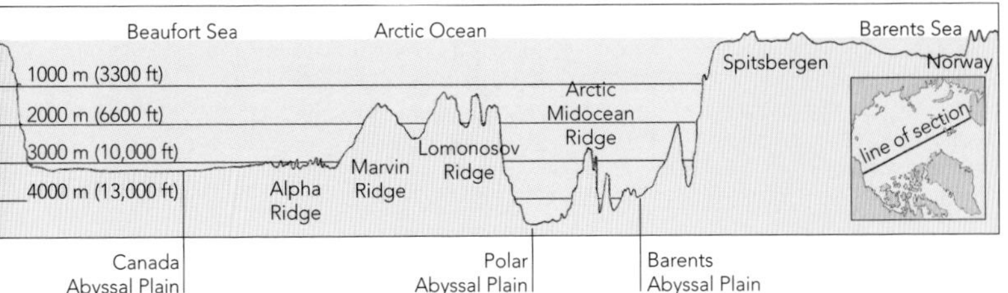

miles (480 kilometers) wide. This extensive shelf area is divided by small island chains into the Chukchi, East Siberian and Laptev shallow sea areas.

▲ ***A cross-section*** *of the Arctic Basin shows the vast differences in height of the ocean ridges.*

Arctic circulation, icebergs and sea ice

Compared to the world's other oceans, the Arctic Ocean is virtually enclosed. The vast majority of water movement, about 80 percent, flows via the Greenland Sea through a narrow gap between Greenland and Svalbard, the only deep-water connection to the world ocean. The remaining 20 percent passes through the shallow Bering Straits into the Pacific.

Two percent of the water leaving the Arctic Ocean does so in the form of icebergs, the majority of which are of the 'glacier' type, being calved from the Greenland icecap and carried into the North Atlantic by the Labrador Current. Such icebergs usually stand about 260 feet (80 meters) out of the water and are more than 3300 feet (1000 meters) long. A second type of iceberg comes only from the ice shelves at North Ellesmere Island and North Greenland. Made up of very old ice, 'ice islands' are small, being around 17 feet (5 meters) high and 660 feet (200 meters) long.

Arctic icebergs range in color from an almost pure white, through green-blue shades to browns and blacks; the different colors reflect the amount and type of soil and/or debris that is mixed in with the ice.

During the winter, sea ice extends over an area of 6 million square miles (15 million square kilometers) of the Arctic Ocean, shrinking to around half that during the summer. This ice cover affects the surface circulation in the Arctic by restricting heat exchange between the ocean and the atmosphere. Water directly below the ice is much more variable in temperature and density than is normal for ocean surface waters.

▼ ***Harsh conditions*** *in the North Atlantic are caused by fog, due to humid winds blowing over the meeting of cold and warm currents, and icebergs carried by the Labrador Current.*

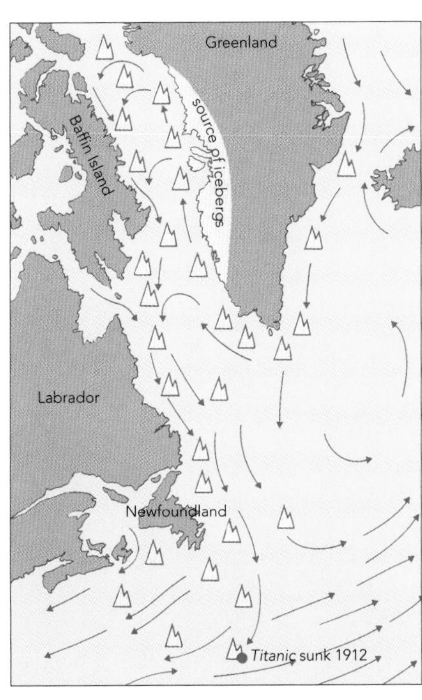

ARCTIC RESOURCES

Despite the inhospitable conditions of the Arctic region, the Arctic Ocean's broad sedimentary basins represent suitable geological formations for the accumulation of oil and gas deposits. Anhydrite deposits are known to occur in the Laptev Sea and off the Canadian coast. In the Beaufort Sea, large oil deposits have been discovered, and gas reserves are extensive in the region of Melville Island.

Oil extraction

Oil was discovered on the Arctic coast of Alaska in 1968, and nine years later, the 800-mile (1270-kilometer) trans-Arctic pipeline was constructed. The pipeline carries crude oil south from Prudhoe Bay on Alaska's Arctic coast. On the Russian coastline, known oil and gas reserves have not yet been exploited.

Problems of oil and gas extraction include the extreme cold, gale force winds and constantly shifting sea ice. In a number of instances, artificial islands have been constructed by dredging and pumping seabed gravel and sand to provide a base for drilling rigs. Land-based pipelines have proved both technically and economically more practical than early ideas concerning the possibility of submarine tankers carrying crude oil under the Arctic ice sheet. Due to the low density of human populations surrounding the Arctic Ocean and the inhospitable conditions, mineral extraction from the Arctic remains at a relatively low level at present.

▼ ***The prow of a ship*** *breaks through an ice-floe near the North Pole. While the ice-breakers' primary role is to keep sea lanes open, they also undertake valuable scientific research in the Arctic.*

Living resources

The living resources of the Arctic depend on a system of primary production which is highly seasonal, with little or no photosynthesis occurring during the winter. Total primary production is only around one tenth of that in temperate ocean areas. During February, when light reaches the Arctic, small phytoplankton, mainly diatoms, begin to increase their productivity. Located on the undersurface of the ice and in areas of open water, by March they have formed a yellow-brown layer within the bottom 12 inches (30 centimeters) of ice. The plankton are fed on directly by the Arctic cod, the only commercially important species found in the central Arctic Basin. On the shallow continental shelves, the fisheries are dominated by Arctic char and capelin.

There are two main communities of fish in the Arctic. First, the cod, haddock and plaice which inhabit the warmer waters of the West Spitzbergen and North Cape currents, and second, the Arctic cod and capelin, which inhabit colder water areas. In response to long-term climatic fluctuations, the distribution and abundance of these two communities of fish alter, reflecting changes in the current systems and ocean circulation.

▲ ***An Inuit hunter*** *slices a dead narwhal. The Inuit of the far north have a close relationship with their environment, taking enough to provide them with food and clothing, but never overexploiting.*

Fishing and hunting

Although cod have been caught on their spawning grounds around the Lofoten Islands since the 12th century, they were not fished from the nearby Barents Sea until the 1920s, when the temperature of the North Atlantic reached a peak. During the 1960s and 1970s, capelin became more important in Arctic catches, the shift reflecting a period of colder climate.

By the early 1970s, rising fuel costs and restrictions placed by the Canadians on the North Atlantic fishery resulted in an increase in fishing effort in the Barents Sea, such that by 1975, a catch quota of 810,000 tons was imposed by the Northeast Atlantic Fisheries Commission on the Arctic-Norwegian cod stocks. All the major fisheries of the Barents Sea are seasonal, with cod, haddock and redfish being taken between February and September, and capelin being fished over the winter months.

The Arctic attracted commercial hunting expeditions, particularly for fur seals, from as early as the 16th century. Commercial hunting of marine mammals has now been stopped, although the indigenous peoples of North America, Siberia and Greenland continue to hunt marine mammals and birds for subsistence use. The Inuit of the far north take only what they need, and continue to have little impact on the species in the region.

Area: *222,008 sq miles (575,000 sq km)*
Volume: *12,955 cu miles (54,000 cu km)*
Average depth: *308 ft (94 m)*
Maximum depth: *(in Skagerrak) 2297 ft (700 m)*

The North Sea is a large ocean formed by the inundation of an extensive area of the continental crust. This basin has been subsiding throughout its long history, and the rather featureless bottom topography conceals an extremely deep sedimentary basin containing more than 20,000 feet (6100 meters) of sediments deposited over the course of the last 250 million years.

Running close to the coast of Norway lies the Norwegian Trough extending from the North Atlantic to the mouth of the Skagerrak in northern Denmark, where it reaches a depth of 2300 feet (700 meters). A much deeper trough lies hidden beneath the sedimentary deposits, and it is in this trough that the vast majority of North Sea oil and gas deposits are found. A line of deep ocean troughs runs roughly north to south to the west of Ireland, marking the edge of the European continental shelf.

Past ice ages

During the ice ages, ice sheets covered much of the British Isles and flowed into the surrounding basins, extending as far south as the Celtic Sea in the west and the Thames Estuary in the east. Additionally, ice sheets extended from the Scandinavian land mass into what is now the North Sea. The well-known 'Banks' of the North Sea, such as Dogger, Fisher and Jutland banks, are in fact glacial moraines – piles of boulder clay pushed out in front of extending ice sheets and subsequently sorted by the action of bottom currents.

Sand is constantly being moved along the seabed by tidal currents, particularly in areas where strong currents coincide with large storm waves. The approaches to the ports of Liverpool, London and Hamburg are among the many areas where sand has accumulated. Dredging to remove the sand and allow access to shipping has, over the years, cost millions of dollars.

The strait of Dover is also an area of sand deposition with long sandbanks being deposited parallel to the strong currents which flow through the straits. The Goodwin Sands and Norfolk Banks are a hazard for shipping, and the shipping lanes to the major ports of Europe require regular surveying since the sand is constantly shifting.

▼ ***Current flow*** *in the North Sea is generated by a combination of the tides, the prevailing winds, and density differences in the water masses. The pattern varies considerably on a local and seasonal basis.*

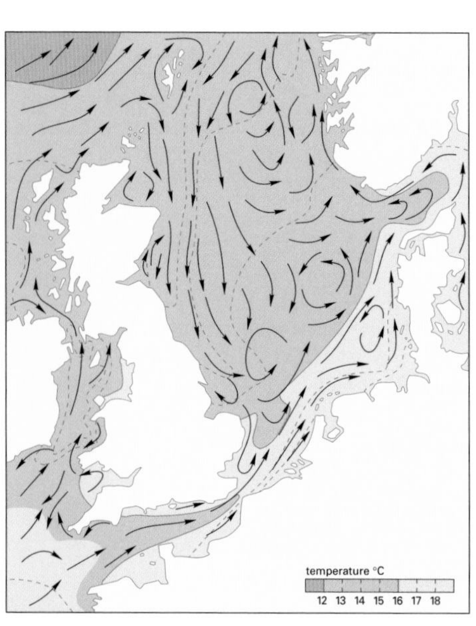

Current circulation

Water movement in the North Sea is generated by the tides, the prevailing winds and density differences between different water masses. In general, the pattern is highly variable both locally and on a seasonal basis. Water enters the North Sea from the north between the Orkney and Shetland Islands, and flows down the eastern coast of Scotland and England. At the same time, warmer water from the North Atlantic Current flows into the North Sea through the English Channel (French, La Manche). The inflow of warm water ensures that the area remains ice free throughout the year. The warm-water current flows along the coast of the Netherlands to the south of the North Sea. The combination of the two prevailing currents sets up a generally anticlockwise circulation in the North Sea.

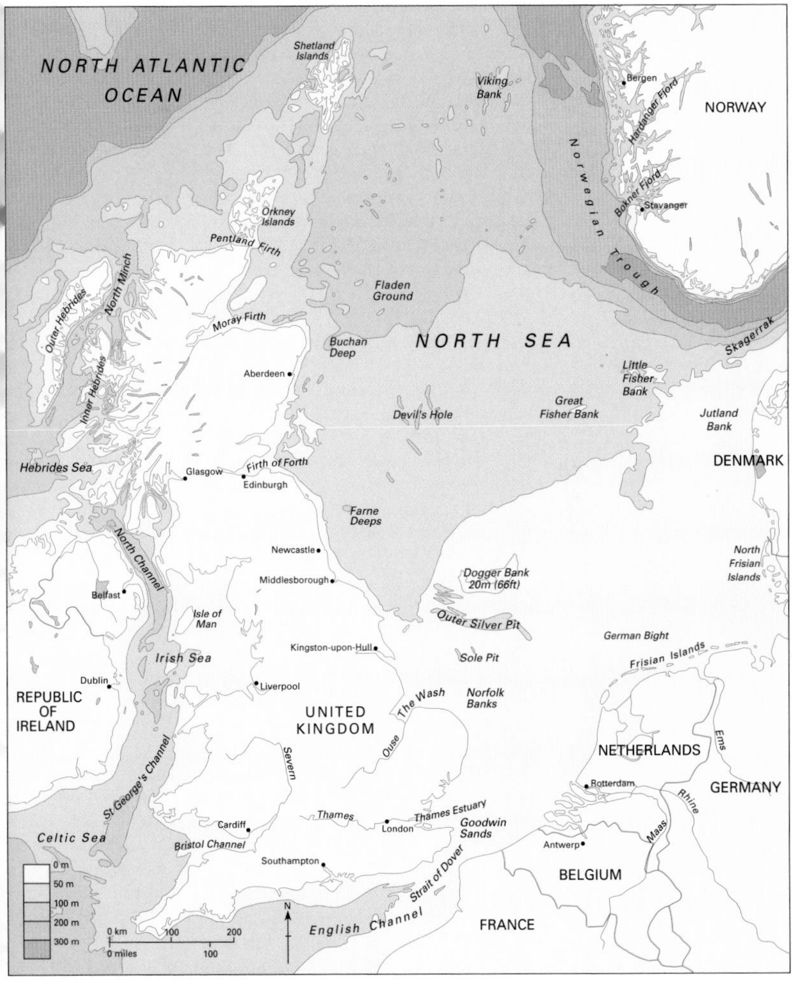

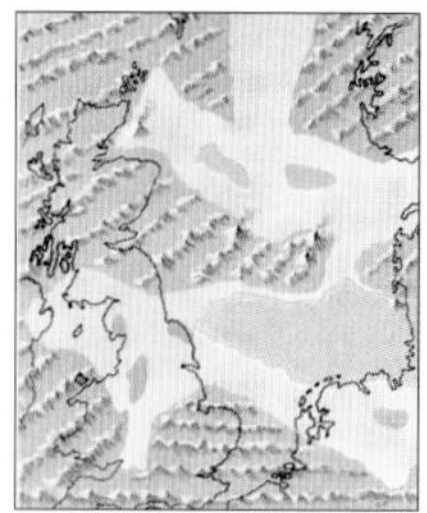

▲ ***During the Permian*** *period, 250 million years ago, the North Sea was a desert plain bounded by mountains. Inland seas and salt lakes were present, forming sandstones.*

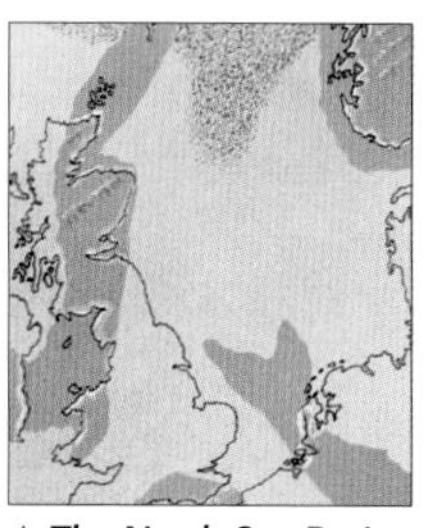

▲ ***The North Sea Basin*** *lowlands were flooded by shallow seas during the upper Cretaceous period, 100 million years ago. Thick chalk deposits were laid down.*

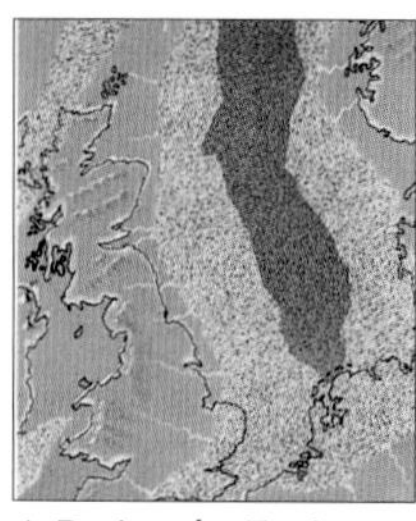

▲ ***During the Tertiary*** *period, the North Sea Basin looked very similar to how it does today and thick deposits of sediments like muds and clays were deposited.*

NORTH SEA RESOURCES

North Sea oil and gas extraction has grown considerably since the early 1970s. Oil production is greatest in the Shetland Basin, where water depths are up to 660 feet (200 meters). Gas extraction from the Southern Bight, on the other hand, occurs in shallower waters at depths of less than 170 feet (50 meters). Although the industry represents a vast economic resource, its extraction brings environmental problems. Recent data suggest that the benthic communities of the North Sea display increasing diversity away from oil rigs, reflecting the influence of low levels of chronic pollution.

Sand and gravel

More sand and gravel is extracted annually from deposits in the North Sea than anywhere else in the world. Although the large quantities extracted reflect, in part, depletion of land-based sources, the materials recovered are well sorted, uniform and of high quality, requiring little processing before use. Most of the dredgers operating in the North Sea work in water that is less than 115 feet (35 meters) deep and close to shore, often resulting in coastal erosion of neighboring land. Gravels dredged off the British coast contain a high proportion of flint and quartzite derived from the weathering of the Cretaceous chalks during the Tertiary period (65 to 2 million years ago).

Living resources

Fishing in the North Sea dates back to the Dark Ages between AD 500 and 1000. The major component of the catch was herring, which were caught by drift nets at night. After the Napoleonic Wars and the growth of urban populations in Western Europe, fishing intensity greatly increased, with the first steam trawler being launched in 1881. More recently, the introduction of purse seines and midwater trawls increased the fishing pressure to such an extent that North Sea herring has been virtually exhausted. In 1977 a total ban on herring fishing in the North Sea was introduced.

Of the flatfish caught in the North Sea, plaice are the most important. The major spawning area is in the south, off the coasts of Holland and Belgium. The eggs drift with the current a few kilometers each day toward Denmark. Although the southern spawning area is the most important, spawning also occurs off the east coast of England, between Flamborough Head and the Dogger Bank, and in the German Bight area.

Overfishing

The North Sea fishery has shown a number of major changes in the last few decades, with catches of cod rising steadily since 1955 and declining alarmingly in the 2000s, and very large catches of haddock occurring in the 1960s. Industrial fisheries (those species used for fish meal, such as Norway pout), also increased during the same period, while catches of herring and mackerel halved. The high intensity of fishing has affected fish stocks in the area, but fluctuations in stocks also reflect changes in oceanic conditions and circulation patterns. Although the European Union introduced a quota system, based on the total allowable fish catch, in order to reverse stock depletion, many commercial species, particularly mackerel, haddock, whiting, saithe and Norway pout, are still in decline.

Farming of Atlantic salmon is increasingly important and is practiced in Scottish sea lochs and Scandinavian fjords. Salmon eggs are incubated in trays and one-year-old smolts are acclimatized to natural conditions over three weeks before being put into the sea in net cages. Natural water movement oxygenates the water and removes wastes, but high-intensity fish farming can cause environmental problems, such as anoxia (depletion of oxygen) of the bottom water.

▼ ***This map*** *of the North Sea shows the region's most valuable concentrations of mineral resources.*

THE MEDITERRANEAN SEA

The Mediterranean Sea
Area: *1,145,000 sq miles (2,966,000 sq km)*
Average depth: *4921 ft (1500 m)*
Maximum depth: *(Hellenic Trough) 16,706 ft (5092 m)*
Evaporation: *roughly three times precipitation and run-off*

The Mediterranean Sea lies in a 2500-mile (4000-kilometer) depression running from the coast of Israel, Lebanon and Syria in the east to the narrow Strait of Gibraltar in the west. Although it covers an area of around 1 million square miles (2.5 million square kilometers), this inland sea is much shallower than most ocean areas.

The two basins

The narrow Strait of Sicily divides the Mediterranean into two distinct basins. The western basin has a broad, smooth abyssal plain that is covered with sediments dating from around 25 million years ago. In contrast, the eastern basin is divided by the Mediterranean ridge system of folded and uplifted sediments compressed by the movement of Africa northward toward Eurasia. Sediments in this basin date back to 70 million years, and the deepest parts of the modern Mediterranean Sea are found here, reaching 16,000 feet (5000 meters) deep.

In the western basin, the Balearic and Tyrrhenian subbasins are separated by the island chain of Corsica and Sardinia. In contrast to the relatively flat profile of the bottom topography of the Balearic Basin, the Tyrrhenian subbasin, which reaches depths of 12,000 feet (3600 meters), is dominated by ridges, seamounts and an arc of active volcanoes including Vesuvius, Etna and Stromboli.

Man-made connections

In addition to its natural connection with the Atlantic via the Strait of Gibraltar, the eastern Mediterranean Basin connects

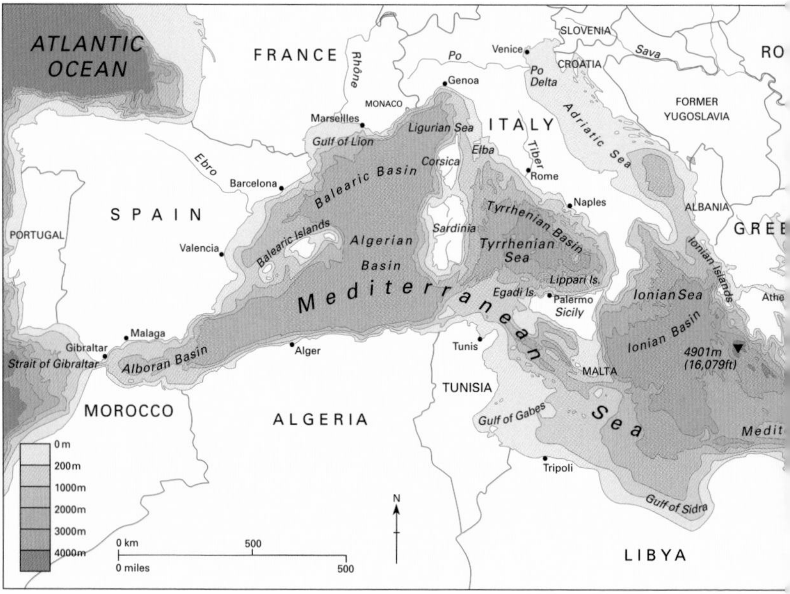

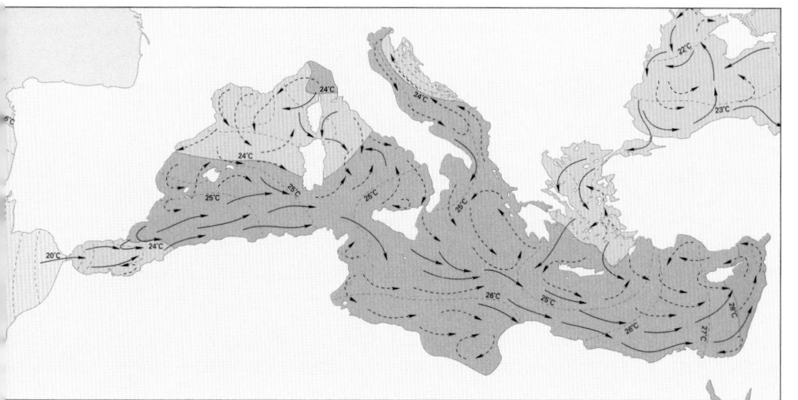

▲ ***Surface water*** *generally flows into the Mediterranean from the Atlantic, while at depths of around 280 ft (80 m) the flow is reversed leaving the Mediterranean via the same route.*

with the Indian Ocean via the Suez Canal, which opened in 1869. Although water exchange via the canal is insignificant, a number of Red Sea species have invaded the Mediterranean Basin via the canal.

The inland Black Sea connects to the eastern Mediterranean Basin via the Dardanelles, the Sea of Marmara and the Aegean Sea, and plankton studies have revealed that zooplankton species characteristic of the Aegean are spreading into the Black Sea. This may result from the damming of rivers which feed the Black Sea, thus reducing the flow of water out through the Dardanelles and increasing the influx of Aegean water.

Current circulation

Although Atlantic surface water enters the Mediterranean through the Straits of Gibraltar, there is a significant outflow of more saline, denser Mediterranean water via the same route. According to legend, this deep-water outflow was harnessed by the Phoenicians, who lowered their sails several fathoms into the water, to harness this current in order to enter the Atlantic against the prevailing winds. Today, this strong undercurrent is used by submarines, which can turn their engines off and pass silently through the straits into the Atlantic.

The Atlantic water entering the basin passes eastward, becoming progressively more saline as water vapor evaporates from the sea surface. Salinity may reach as much as 39.5 parts per thousand in the eastern Mediterranean. In the summer, a marked thermocline at around 66–130 feet (20–40 meters) separates the warmer, high-salinity surface water from the cooler water below. During winter, dry winds also remove water vapor, increasing surface salinities and at the same time cooling the surface water, which becomes denser and sinks. These winter water masses flow westward toward the Strait of Gibraltar, or sink to comprise the dense bottom water of the eastern basin.

The Mediterranean loses by evaporation almost three times as much water as it receives from rainfall and run-off from the surrounding land. This imbalance is compensated for by the inflow of Atlantic water, which flows eastward, hugging the North African coast, creating the only well-defined current in the Mediterranean. This current feeds the anticlockwise circulation patterns of the western basin, the Adriatic and Ionian Seas. Water exchange with the Atlantic is limited, and the turnover time of the Mediterranean is estimated at 150 years. Tidal range in the Mediterranean is also low, approximately 12 inches (30 centimeters).

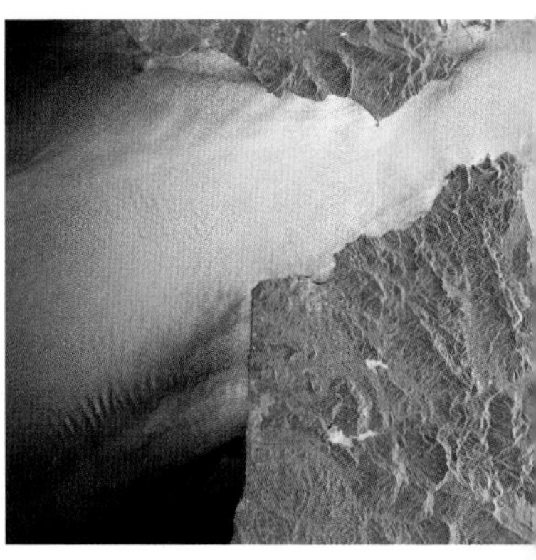

▲ ***This satellite picture*** *shows waves at the Strait of Gibraltar produced by tidal flow through the narrow strait.*

MEDITERRANEAN RESOURCES

Seismic studies of the Mediterranean record the presence of a widespread stratum of evaporites, which may be as much as 3000 feet (915 meters) thick and is conservatively estimated at 240,000 cubic miles (1 million cubic kilometers) in volume. This represents a substantial potential resource of rock salt, sulfur and potash, all of which are currently only exploited where they outcrop on Sicily and other Mediterranean islands. The Mediterranean region also has considerable oil and gas reserves, but to date these are only exploited in offshore shelf areas that overlie geological structures extending out from the land. Manganese- and iron-rich deposits exist in metalliferous muds derived from the recently active submarine vents in the Eastern Basin.

Tourism

Perhaps one of the most valuable resources of this area is the combination of warm dry summers and the Mediterranean Sea itself. Together, they provide the basis for an extensive tourist industry. With more than 30 percent of the world's tourists visiting the Mediterranean region each year, tourism contributes significantly to the economy of Mediterranean countries.

▼ ***A crowded beach*** *in the popular French resort of Cannes. Tourism is a major source of revenue to the countries surrounding the Mediterranean.*

Living resources

The Mediterranean contains some 500 species of fish, of which around 120 are fished commercially. The high market price and high seasonal demand resulting from the tens of millions of tourists who visit the Mediterranean each year, encourage overfishing. The fisheries which are most heavily exploited are those along the southern coast of Europe, where there are severely depleted stocks of hake, sole and red mullet. The present harvest of more than 2 million tons of fish per annum is well in excess of the estimated sustainable yield of 1.1–1.4 million tons.

In general, phytoplankton production in the Mediterranean is limited. The exchange of nutrients from the cold, bottom

▲ ***Sicilian fishermen*** *pull a bluefin tuna into their boat during a* mattanza. Mattanzas *occur each year when the tuna return to reproduce.*

water to the surface is limited, and there are few rivers bringing nutrients into the sea. The construction of the Aswan Dam and various barrages on the Nile River have significantly reduced the input of nutrients into the eastern basin and may have contributed to the collapse of the sardine fishery in the delta region. In addition, while the fisheries along the North African coastline are not yet overexploited, fishing intensity is also increasing in this area.

Mariculture

Mariculture in the area has considerable potential, with up to 4000 square miles (10,000 square kilometers) of coastal lagoons providing considerable potential for future mariculture of finfish and shellfish. Today, the most important species in the Mediterranean is the Mediterranean mussel, principally produced by Italy. Other important species include the mullet, gilt head bream and sea bass. The culture of species such as sea bream in cages suspended from anchored vessels, is being tried experimentally, and large sections of the Adriatic coastline are now lined with net cages suspended from flotation rings.

The Indian Ocean
Area: *31,660,000 sq miles (73,600,000 sq km)*
Volume: *70,086,000 cu miles (292,131,000 cu km)*
Average depth: *12,760 ft (3890 m)*
Maximum depth: *(Java Trench) 24,442 ft (7450 m)*

The Indian Ocean is the world's third largest ocean (afte the Pacific and Atlantic) and represents approximately 2C percent of the total area of the world's oceans. The floor o the Indian Ocean is dominated by the midocean ridge, which forms an inverted Y-shape, the western arm of which runs round the southern tip of Africa to join the Mid-Atlantic Ridge. The eastern arm passes south of Australia to connect with the East Pacific Rise.

The surface ocean conditions of the Indian Ocean are dominated by the monsoon winds. However, this ocean basin is open to the south and connects with the Southern Ocean, such that cold, deep Antarctic waters penetrate a considerable distance to the north in the deeper ocean basins. Storms generated in the turbulent atmosphere of high latitudes result in long-distance swell transmission northward into the western Indian Ocean, causing periodic flooding in island countries such as Sri Lanka and the Maldives.

Geological history

The formation of the Indian Ocean began with the breakup of Gondwanaland, when the African continent separatec

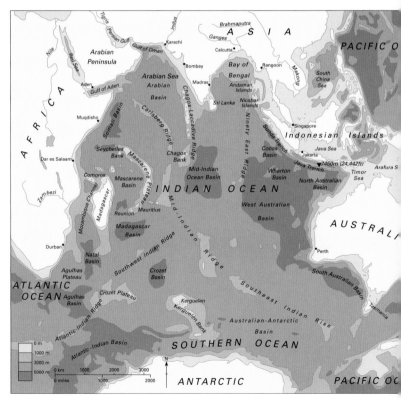

from Antarctica and Australia between 140 and 130 million years ago. Seventy million years ago, India lay south of the equator, but moved northward to collide with the Eurasian continental plate, resulting in the formation of the Hindu Kush and the Himalayas. In its present form, the Indian Ocean basin is relatively young, a mere 36 million years old.

Some of the bottom features, notably the Ninety East Ridge which is more than 1700 miles (2720 kilometers) long, were originally at, or near, sea level. The ridge appears to have formed from a single volcanic source, and at its northern and oldest end, the ridge is more than 1 mile (1.5 kilometers) below sea level. In addition, several of the relatively shallow submarine plateaus appear to be continental fragments left behind during sea-floor spreading, and which have subsided as the larger continental land masses drifted apart.

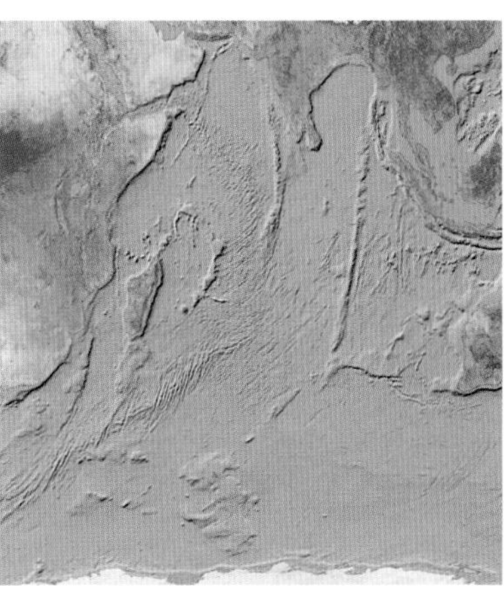

▲ ***This false-color*** *image indicates the position of the Indian Ocean trenches.*

Atoll groups

One of the features of the Indian Ocean, like the Pacific, is the presence of atolls, ringlike structures of living coral reef – indeed the word 'atoll' comes from the Indian Ocean, being the Maldivian word for 'place'. The living reef supports small sandy islets and grows on the consolidated limestone, laid down by previous reef communities, which lies on top of numerous submarine mountain ranges and volcanoes. The major atoll island groups of the Indian Ocean include the Maldives and the Seychelles, and in the case of the Maldives, the formation of the atolls differs considerably from those atolls found in the Pacific Ocean region.

Unlike Pacific atolls, the atoll lagoons of the Maldives contain a wide variety of structures including faros, micro-atolls, and patch reefs. Faros are mini-atolls, small ring-shaped structures often supporting a small island and with a shallow lagoon fringed by living coral reef. The lagoon of a faro is often shallower than that of the main lagoon, and many of the island tourist resorts, for which the Maldives is well known, have been constructed on islands sitting on faros inside the shelter of the main atoll formation.

Bottom sediments

The first sediments deposited on the Indian Ocean seafloor were peats and low-grade coals, characteristic of shallow-water conditions. The shallow submarine plateaus are generally covered by submarine calcareous oozes, while the deeper ocean basins contain mainly reddish-brown clays or silicaceous oozes.

Two of the world's largest river systems, the Indus and Ganges-Brahmaputra, empty into the Indian Ocean and have built enormous fans of sediment brought down from erosion of the Himalayas. The Bengal fan is the world's largest, containing an estimated volume of 5 million cubic kilometers (1.2 million cubic miles) of sediments. The large volumes of freshwater entering the Bay of Bengal result in lowered surface salinity, around 34 parts per thousand, compared with 36 parts per thousand in the Arabian Sea.

The Indowest Pacific

The Indian Ocean covers an enormous area and is connected to the east with the Pacific Ocean Basin via the southeast Asian archipelagos. As a tropical ocean area, it is a center of marine biodiversity with numerous species of coral, fish and other marine organism being found in the shallow waters of the East African coast and peninsula India. In general terms, the diversity of organisms decreases north and south of the equator, although more than 200 species of coral are recorded

Water profiles of the Indian Ocean

The Antarctic Cirumpolar Current dominates the surface of the southern Indian Ocean, sweeping west to east and completing the Southern Indian Gyre. Below the gyre, cold Antarctic Bottom Water sinks and spreads out at depth, also circulating west to east but

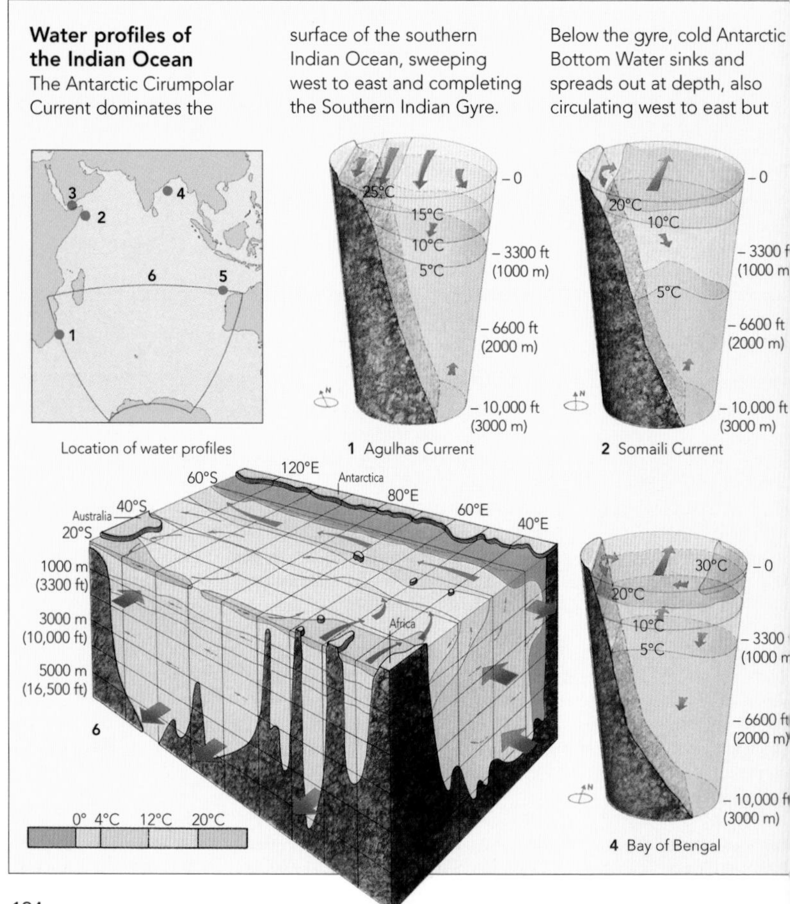

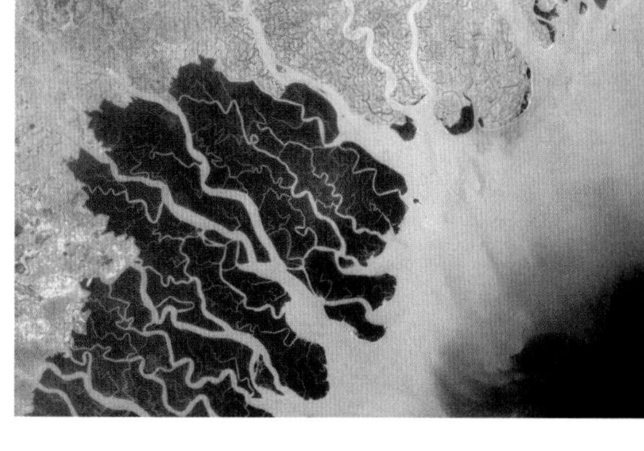

► ***Satellite image*** *of the Bay of Bengal. As the Ganges and Brahmaputra rivers empty, they carry large quantities of sediment into the bay. In this picture, the sediment is seen as a muddy color in the otherwise blue Indian Ocean. The dark part of the delta is the Sundarbans, a wildlife preserve and mangrove swamp, habitat of the Bengal tiger.*

nding to the equator
e bottom of the
p West Australian and
-Indian Ocean basins.

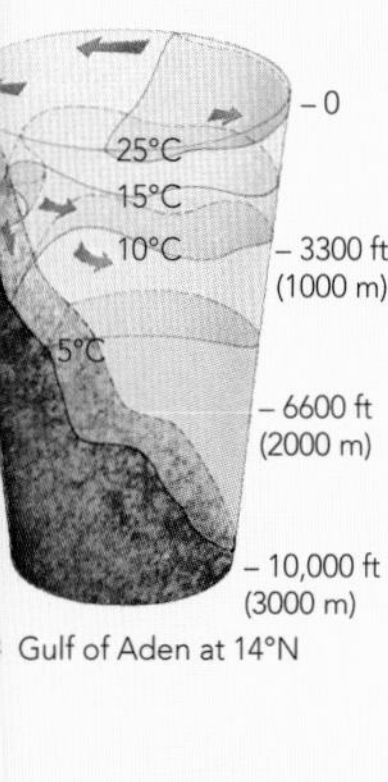

Gulf of Aden at 14°N

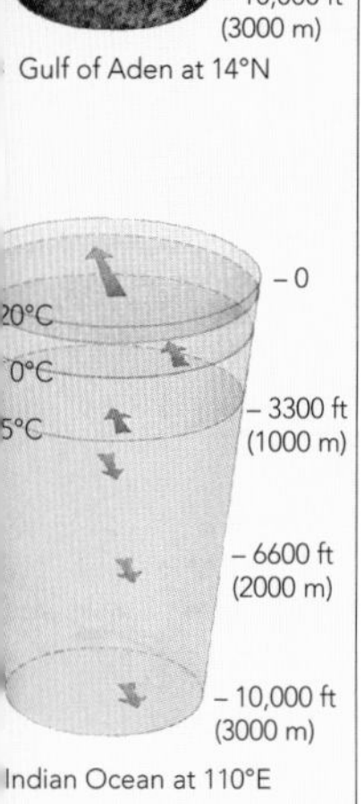

Indian Ocean at 110°E

from the North and Central Red Sea. The relative isolation of the tropical Indian Ocean from the Pacific and its total isolation from the tropical Atlantic has allowed the evolution of distinct shallow-water species in the western Indian Ocean Basin. Similar isolation of the central Pacific has resulted in the evolution of endemic species in the Pacific Ocean, as well. Between these two ocean basins lies the Indowest Pacific, a biogeographic region with the highest shallow water marine species biodiversity in the world, having accumulated species from both the central Pacific to the east and the Indian Ocean to the west.

THE INDIAN OCEAN BASIN

From the junction of the two midocean ridges, which link the Indian Ocean Ridge with the Atlantic and Pacific ridge systems, the midocean ridge runs northward before swinging west as the Carlsberg Ridge to join the rift system of the Red Sea. At its point of entry into the Red Sea, Africa and the Arabian Peninsula have been actively spreading apart for the last 25 million years.

Off the southeastern African coast lie the Agulhas Plateau, Mozambique Ridge, and Madagascar and Mascarene plateaus. These are aseismic structures supporting islands and presumed to be fragments of continental crust. The Chagos Laccadive Ridge off the west coast of India is a stable ridge supporting an extensive chain of atoll islands. To the east of this ridge, the seafloor drops to the Chagos Trench. Further east still, lying south of the Indonesian island arc, is the only major trench in the Indian Ocean – the Java Trench. This trench reaches depths of 24,400 feet (7300 meters) and marks the subduction zone where the Australian plate slides below the Eurasian plate, a process that has been active for only 2 million years.

Related volcanic activity in the Indonesian region has resulted in the build up of extensive deposits of ash and volcanic sediments on the seafloor in this area. The well-known eruption of Krakatoa in 1883 resulted in about 4 cubic miles (17 cubic kilometers) of mountainous island erupting in a series of four explosions, the loudest of which was heard in Australia 3000 miles (4800 kilometers) away.

North Equatorial Current

The northern Indian Ocean circulation is unique in that its surface currents reverse twice a year under the influence of the monsoon winds. From November to April, the northeast monsoon generates the North Equatorial Current. This current carries water across the Indian Ocean toward the African coast. Here the current turns south, forming a western boundary current that flows along the coast of Somalia toward the equator. At the equator, it joins the South Equatorial Current and flows east as the Equatorial Countercurrent. In the region of the Indonesian island arc, this current divides, one part flowing north to rejoin the North Equatorial Current and the other continuing to flow eastward as the Java Coastal Current.

In April, the northeast monsoon ceases and is replaced by the southwest monsoon winds, which generate a north-flowing current off the Somali coast. By July, the monsoon current flows east. After reaching the Indonesian islands, the entire water mass turns south and returns westward as part of a much stronger South Equatorial Current, which in turn feeds the Somali current.

During the southwest monsoon, not only are the currents stronger, but significant areas of coastal upwelling occur off the Arabian Peninsula and Somalia where warm surface water is replaced inshore by upwelled colder water from between 330 and 660 feet (100 and 200 meters).

▶ ***The currents*** *in the northern Indian Ocean are reversed according to the monsoon seasons. During the northeast monsoon, the surface waters are driven away from India toward Africa, and the Somali upwelling is suppressed. Water from the Arabian Sea enters the Gulf and Red Sea during this monsoon season.*

South Equatorial Current

South of the equator, the Indian Ocean circulation is dominated by the influence of the westerlies and trade winds, which drive

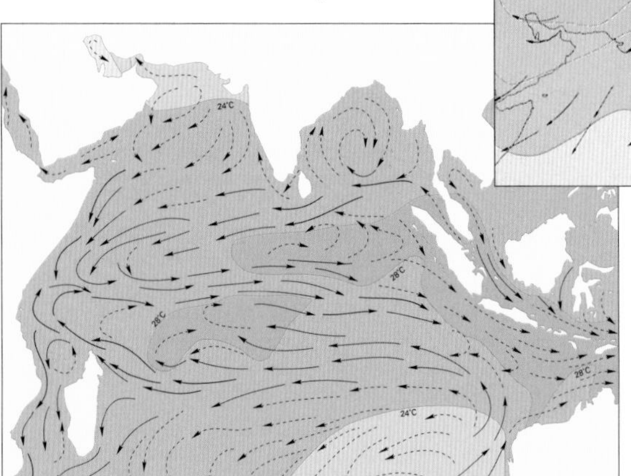

◀▲ ***Northeast Monsoon*** *Strong high pressure over northern Asia causes an overall outflow of air that drives the surface currents to the southwest.*

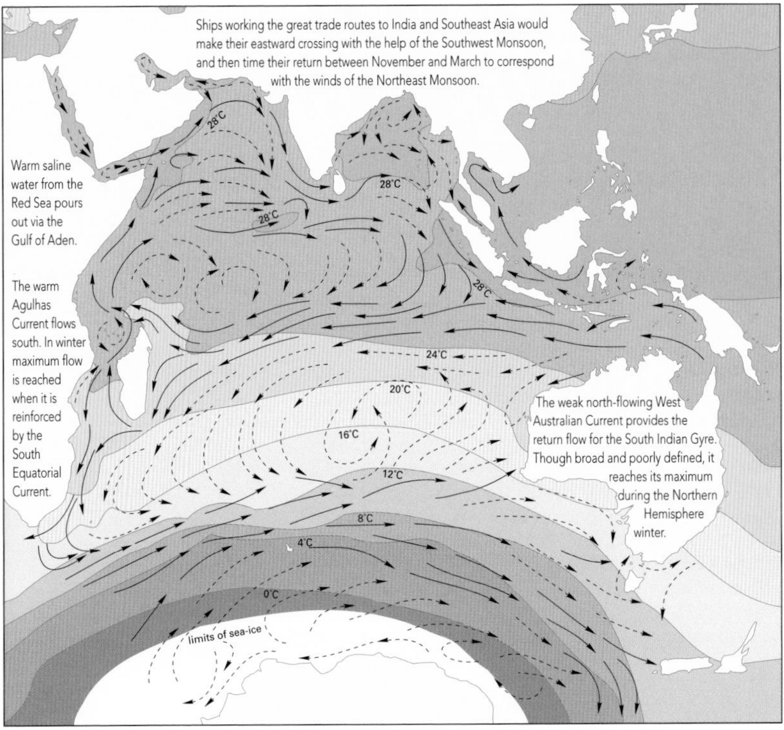

Ships working the great trade routes to India and Southeast Asia would make their eastward crossing with the help of the Southwest Monsoon, and then time their return between November and March to correspond with the winds of the Northeast Monsoon.

Warm saline water from the Red Sea pours out via the Gulf of Aden.

The warm Agulhas Current flows south. In winter maximum flow is reached when it is reinforced by the South Equatorial Current.

The weak north-flowing West Australian Current provides the return flow for the South Indian Gyre. Though broad and poorly defined, it reaches its maximum during the Northern Hemisphere winter.

28°C

24°C

20°C

16°C

12°C

8°C

4°C

0°C

limits of sea-ice

an anticlockwise gyre of warm water. The South Equatorial Current flows toward Africa, where the part that does not feed the Somali Current passes south between Africa and Madagascar and enters the Agulhas Current. This is the strongest western boundary current in the southern hemisphere and flows at a rate of 110 miles (180 kilometers) per day along the edge of the southern African continental shelf.

On reaching the southern tip of Africa, the Agulhas Current turns eastward and occasionally eddies may be cut off and travel westward into the Atlantic. The return flow of the gyre results from the relatively weak and poorly defined West Australia Current, and no area of upwelling is associated with this return.

Most of the water below 3300 feet (1000 meters) in the Indian Ocean originates either as North Atlantic Deep Water or as Antarctic Bottom Water, while in the North Indian Ocean, warm saline water from the Gulf is found at about 1000 feet (300 meters) and below this even more saline water which originates from the Red Sea.

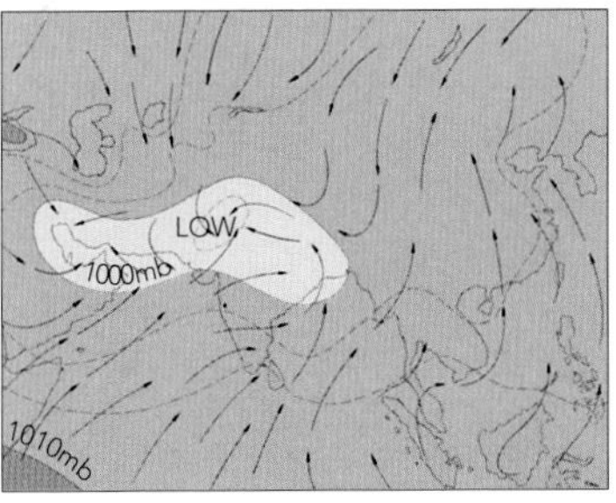

▲ ***Southwest Monsoon***
Heated air rising over the continent in summer creates a low-pressure region. This causes an inflow of air from the south, driving the surface currents.

INDIAN OCEAN RESOURCES

The narrow extent of much of the continental shelf around the Indian Ocean, particularly along the East African coast, has resulted in much lower fisheries production than in either the Atlantic or the Pacific Oceans. The reversal of the surface currents in the North Indian Ocean and the shutting off of the upwelling along the western margins of the basin during the northwest monsoon season, are also thought to contribute to this relatively low productivity.

Mariculture and fisheries

Most capture-fisheries production on the East African coast and around the oceanic island, is from inshore, artisanal fisheries associated with coral-reef areas. These subsistence fisheries are important in providing protein resources to coastal populations in the developing countries surrounding the Indian Ocean basin. As a consequence of the low yield from fisheries production in reef areas, mariculture has been extensively developed in East Asia. The recent growth in mariculture of finfish, shellfish and seaweed, particularly in the Indonesian region, reflects the importance of marine protein to both the subsistence and export sectors of the economy of these countries.

Larger-scale commercial fisheries are either directed toward pelagic resources, such as tuna, sailfish and marlin, or toward penaeid shrimp resources trawled from offshore areas associated with large mangrove stands that support the juvenile shrimps during the early stages of their life cycle. The world's largest stands of mangroves occur in the Sundarbans area, which is found near the Bay of Bengal. Similar large areas in the countries surrounding the South China Sea and in Indonesia have been cleared to provide fuelwood and land for rice cultivation and mariculture.

▼ ***Large cantilever*** *fishing nets at Cochin, southern India. It is believed that traders from the court of the Kublai Khan introduced these nets to the west coast of India in the 13th century. Although the nets are mainly a tourist attraction, fishing is still a vital industry in the southern Indian state of Kerala.*

In the Maldives, the tuna fishery is unique in being based not on large-scale purse seiners or long-line vessels, but on the mechanization and expansion of the traditional *dhoni* fleet. Surface-swimming tunas are caught by pole and line, and transported fresh or on ice to land-based freezer and canning facilities, producing a dolphin-friendly product. Significant large-scale tuna fisheries are currently operating in the western Indian Ocean, based in the Seychelles and Mauritius.

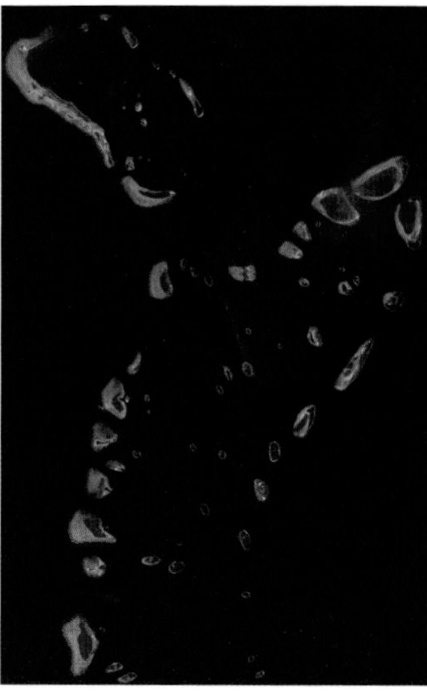

▲ ***Landsat 7 photo*** *of the coral atolls of the Maldives. The Maldives are a group of coral islands resting on top of an ancient volcanic mountain range in the North Equatorial Current of the Indian Ocean off the southwest coast of India. In 2001 nearly 500,000 tourists visited the Maldives, attracted by its pure white beaches and excellent scuba diving opportunities.*

Marine-based tourism

For many of the smaller island nations, tuna and tourism are the sole resources that can be used to generate export income for development. Mass tourism, based on the coral beaches and warm waters of the countries surrounding the Indian Ocean, has grown considerably over the last two decades. The economy of countries dependent on tourism is extremely fragile. During the Gulf War (1991), for example, tourist arrivals in the Maldives dropped to less than half the normal numbers and many resorts went out of business.

An additional problem is the environmental impacts of tourism. In the Maldives, Seychelles and other smaller islands of the Indian Ocean, coral reefs and their associated white-sand beaches form the main attraction for tourists from the northern hemisphere. The damaging impacts of large numbers of people swimming, diving and collecting on coral reefs, has resulted in significant degradation in the environment on which the industry depends.

Nonliving resources

Large areas of the Indian Ocean sea floor contain significant reserves of manganese nodules, particularly in the southern basins. Phosphate nodules of potential value as a source of fertilizers have been located on the Agulhas Plateau.

Placer deposits have been actively exploited around the margins of the Indian Ocean for more than 80 years. Of these deposits, tin (in the form of cassiterite) is perhaps the most significant. This is mined around 5 miles (8 kilometers) offshore from Burma (Myanmar), Thailand and Indonesia. Around a quarter of the world's production of tin came from these deposits during the 1970s. Monazite, ilmenite, rutile and zircon are mined from beach sands along the shores of Kerala state in southern India, while sands rich in ilmenite, rutile, zircon and magnetite are mined in northern Sri Lanka. In eastern South Africa, glauconite, rich in potash, is mined for fertilizer production. Oil and gas production from offshore reserves is still largely confined to the northern Arabian Sea, although significant offshore reserves have recently been exploited in the Indonesian region.

The Red Sea
Area: *169,000 sq miles (438,000 sq km)*
Width: *90–190 miles (145–306 km)*
Maximum depth: *(Axial Trough) 9580 ft (2920 m)*
Evaporation: *more than 80 in/year (200 cm/year)*

The waters surrounding the Arabian Peninsula and its neighboring countries form a distinctive subregion of the northern Indian Ocean. The region includes the Red Sea and the Gulf of Aden to the west and south, the Persian Gulf and Gulf of Oman to the north and east, and the Arabian Sea, which receives high-salinity water from all these sources.

Formation of the Red Sea

The Red Sea represents an ocean in the relatively early stages of formation, and is in reality a rift valley which has been flooded. The Red Sea with its associated arm, the Gulf of Aqaba, is comparatively deep, reaching in excess of 6500 feet (2000 meters) in some areas. In contrast, the Gulf of Suez is quite shallow.

The bottom topography of the Red Sea is dominated by the wide, broad, smooth continental shelf and an axial trough, which is itself split by an even deeper axial valley almost 15 miles (25 kilometers) wide. The axial valley was formed by relatively recent spreading of the seafloor, and the bottom shows volcanic features, including recent lava flows.

On the continental shelf areas, the thick layers of sediment overlie a layer of anhydrite and salt – minerals which are normally associated with evaporite basins of shallow depth. These deposits are around 5 million years old and show that at that time the Red Sea was an area of high evaporation.

Seafloor spreading

On the basis of this evidence, it seems likely, therefore, that two periods of seafloor spreading have occurred. The first was between 20 and 30 million years ago, resulting in the formation of the present continental shelves, and this was

▶ ***This satellite picture*** *shows the Gulf of Suez and Gulf of Aqaba separated by the mountainous Sinai Peninsula. These two gulfs appear to have been formed during different periods of spreading of the Red Sea.*

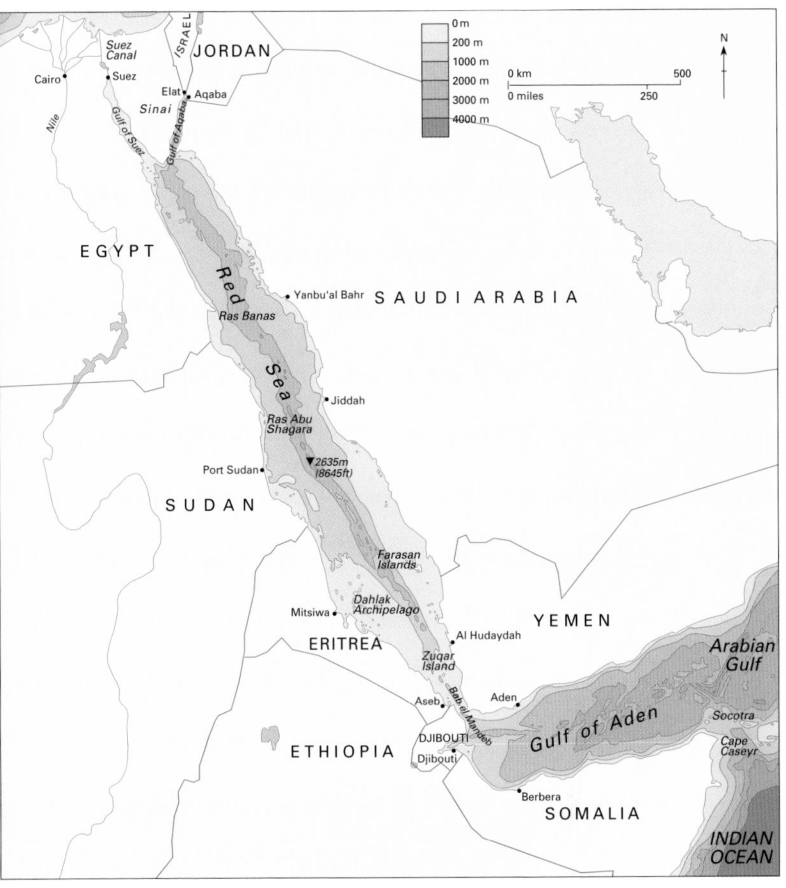

succeeded by an inactive period during which sediments were deposited and an evaporite basin was formed. The second, more recent onset of seafloor spreading commenced around 2 million years ago and has resulted in the formation of the axial valley and the midocean ridge.

Currently, the Red Sea is widening at a rate of about 0.5 inches (1.25 centimeters) per year. Although it is today barely 200 miles (320 kilometers) wide at its widest point, if seafloor spreading continues at its present rate, in 200 million years the Red Sea would be the same width as the present-day Atlantic Ocean.

The uniform thickness of the sediment layers above the evaporite layers suggests that spreading was not occurring. If seafloor spreading had been occurring continuously, the sediments would have been thicker toward the margin, decreasing in thickness toward the axial valley.

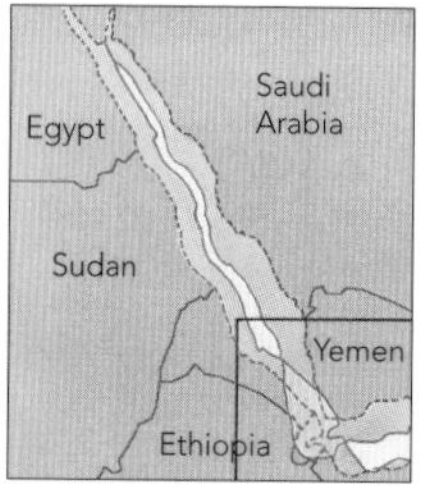

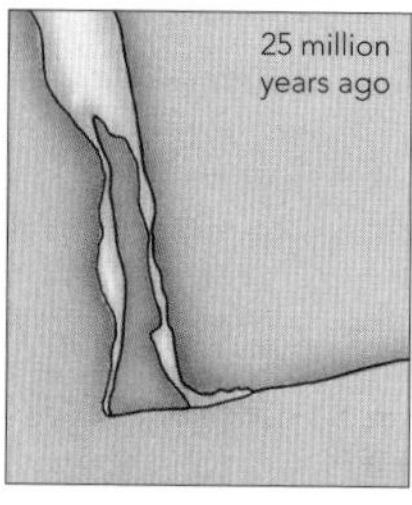

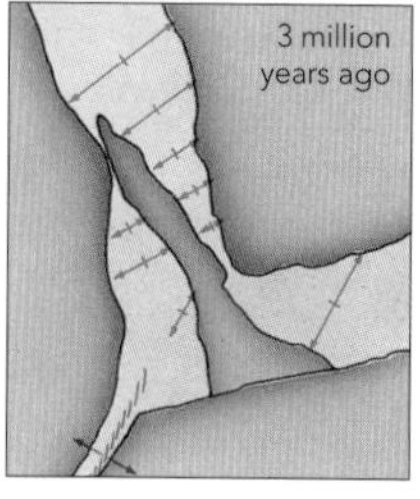

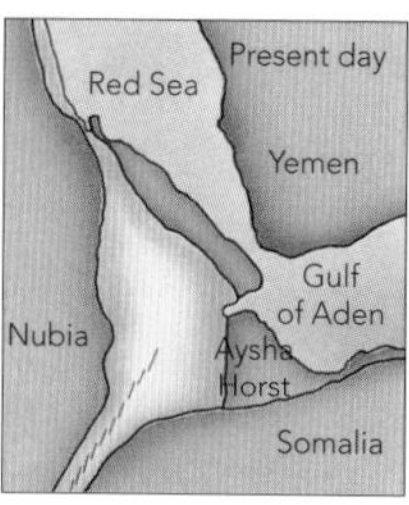

▲ ***About 25 million*** *years ago, the Yemen lay between Somalia and Nubia, which started moving apart from centers at either side of a crust which now forms the Danakil Horst. The area to the southwest became the Afar Triangle.*

The Afar Triangle

The Afar Triangle represents an unusual piece of coastal geomorphology. It is found at the triple-point junction of the Red Sea, the Gulf of Aden and the East African Rift systems. About 25 million years ago, the tip of the Arabian Peninsula sat snugly between Nubia and Somalia. However, two centers of spreading developed on either side of what is now the Danakil Rise and, following uplift, the oceanic crust of the Afar Triangle became dry land. The land surface of the Afar Triangle is inhospitable desert, covered in many areas by thick deposits of salt. Its unique surface features of old, volcanic cones faced with shattered volcanic glass and thick deposits of salts, are seen nowhere else on Earth.

Current circulation

The waters of the Red Sea are relatively clear, and the coastline is fringed with extensive coral-reef systems. Circulation in the Red Sea basin is generally wind driven, and during the northwest monsoon, the flow is toward the Gulf of Suez. During the southeast monsoon, however, the current reverses, passing down toward the narrow strait of Bab el Mandab. Exchange of water through these straits with the neighboring Arabian Sea is dominated by tidal currents.

The reversal of currents is similar to that in the Arabian Sea, resulting in upwelling, which occurs off the coast of Somalia and Oman under the influence of the southwest monsoon winds. These pass parallel to the coast and cause the surface waters to flow away from the coast, drawing up colder, nutrient-rich waters from beneath. During the northeast monsoon season, the cold waters become trapped below the warmer surface waters and the subsequent lowered nutrient supply results in reduced primary production.

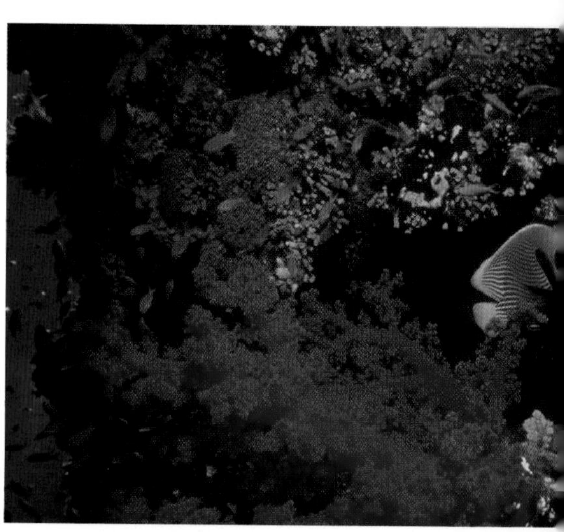

RED SEA RESOURCES

The clear waters of the Red Sea and low, fresh-water run-off from the surrounding arid lands, have resulted in the development of extensive coral-reef systems along both shores. The coral reefs extend for a distance of nearly 1250 miles (2000 kilometers) along the shores and dominate the marine environment of the area.

More than 350 species of corals have been recorded in the Red Sea region, which has a much higher diversity of marine organisms than either the Persian Gulf or the Arabian Sea. The diversity of corals is greater even than that of the Caribbean Sea region and equal to the highest recorded in the Indian Ocean as a whole. An estimated 6 percent of the species are found only in the Red Sea, and as many as 90 percent of some groups of coral-reef fish are unique to the area. Unlike other tropical seas, mangroves are not well developed in the Red Sea.

The diverse fish fauna includes important food fish, such as snappers, grouper and parrotfish, which are exploited by small-scale artisanal fishing industries. Since the density of human populations along the coastlines of the Red Sea is generally not great, artisanal fishing pressure is commonly low. To the south of the region and in the north around Ghardaqa, small commercial trawl fisheries for demersal species have been established.

The most important fishery of the area is a pelagic fishery for sardine in the northern reaches of the Red Sea and Gulf of Suez. The sardines are caught at night by means of large, circular nets used in conjunction with lights, which attract the fish to the surface above the suspended net.

▼ ***Different varieties*** *of coral in the Red Sea. The diversity of coral species encourages a variety of marine-life to breed in the area. The coral reefs are a vital resource in the Red Sea region.*

Metalliferous muds

Metallic muds and brines were first discovered in the Red Sea region in 1963, and are associated with the volcanic activity of the axial rift valley. The submarine rocks of the region contain significant concentrations of zinc, copper, manganese and lead, and as the saline waters of the Red Sea permeate through the fractured submarine rocks, they leach out the metals. The hot brine, brought to the surface through submarine vents, is trapped in deep pits more than 6000 feet (1850 meters) deep, and as the brine mixes with the cooler Red Sea water, the metals are precipitated out.

Although the metal concentrations may be high, the costs of recovery are such that economic extraction is not possible given current metal prices and available reserves. Of the 15 hot brines discovered in the Red Sea region, the Atlantis II Deep is believed to contain the highest concentrations of metals, thereby making it the most viable in economic terms. The Atlantis II lies within the exclusive economic zone of Sudan and Saudi Arabia, and to control exploitation in the area a Red Sea Commission has been put in place. In contrast to the Persian Gulf, oil and gas reserves are not important in this region.

The Persian Gulf
Area: *93,000 sq miles (240,000 sq km)*
Width: *35–210 miles (56–340 km)*
Maximum depth: *(Strait of Hormuz) 330 ft (100 m)*

In contrast to the Red Sea, the Persian Gulf is extremely shallow, with the majority of its sea floor lying only 100 meters (330 feet) or less below the surface. In addition, unlike the Red Sea, the Persian Gulf represents an area of active subduction where the Arabian Peninsula is sliding beneath the Asian continental plate. As a consequence, the northeastern shore drops rapidly into a trough, while the opposite shore shelves more gently.

The Arabian Peninsula is one of the world's smallest continental plates. Oceanic crust is formed in the Red Sea rift valley and is pushing the Arabian plate northeastward, where it is lost beneath the Asian plate. The Zagros Mountains of Iran, which lie close to the northeastern shore, represent the folded margin of the Asian plate which is being compressed as the Arabian plate slides beneath it. The sediments that are being deformed at the foot of the Zagros Mountains include deposits of salt, pushed up into domes, which give rise to the extensive oil fields of the region. The continued movement

Shatt al Arab
Basra
IRAQ
Abadan
KUWAIT
Bubiyan
Kuwait City
Khark
Bushehr
IRAN
Persian Gulf
Jazireh-ye Lavan
Qeys Island
Dammam
Manama
BAHRAIN
Ras Rakan
Doha
QATAR
SAUDI ARABIA
Abu
UNITED ARAB EMIR
0 m
20 m
60 m
100 m
200 m
0 km 100
0 miles 100
N

of the Arabian plate may ultimately result in the complete closure of the Persian Gulf, a process that could take only a few tens of thousands of years.

Current circulation

Due to the shallow nature of the Persian Gulf basin, the waters are well mixed, and nutrient inputs from the neighboring land mean that primary productivity is higher than in the Red Sea. The diversity of species is much lower than in the Red Sea and the coral reefs are not as well developed. However, the Gulf's stands of mangroves and seagrass meadows are more extensive than in the Red Sea.

Surface circulation is driven by the winds, and warm, high-salinity water passes out of the Gulf into the Arabian Sea during the southwest monsoon. Since the water is not as highly saline as that of the Red Sea, it disperses out into the Arabian Sea at depths of only around 600 feet (185 meters).

The temperatures of water in the Gulf are generally high, reaching as much as 91°F (33°C). The coral in this semi-enclosed basin has adapted to the harsh conditions and is capable of withstanding temperatures which would result in bleaching and mortality on reefs in more open and cooler water environments.

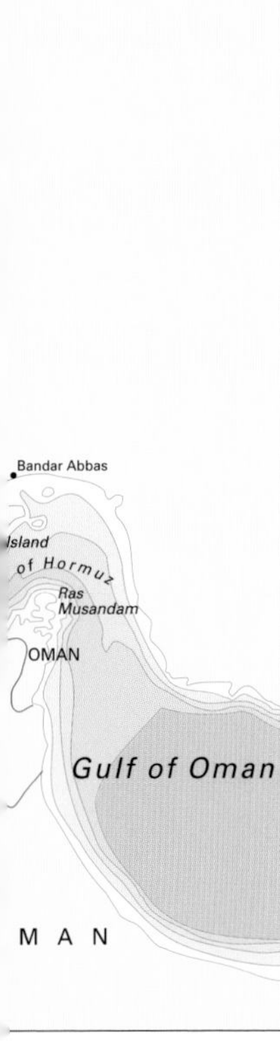

PERSIAN GULF RESOURCES

The Persian Gulf is synonymous with oil, and indeed the mineral resources of this area are both rich and diverse. What is perhaps less well known is that, as a consequence of the inputs of nutrients from the Zagros Mountains and the Gulf's shallow depth which results in well-mixed waters, the biological productivity of this area is also much higher than that of the Red Sea.

Although the region's first oil well was sunk in 500 BC, at Shush in Iran, the history of modern oil exploration and exploitation is a mere 100 years old. Modern exploration began in the 1890s but was relatively unsuccessful until the discovery of the Naft-i-Shah field in Iran in 1923, which commenced production in 1935. Today, this area remains one of the major oil-producing regions in the world, exploiting a number of land-based and offshore fields.

Sedimentary rocks have been deposited in the Persian Gulf region for nearly 280 million years. They consist of a sequence of permeable limestones, interbedded with organic-rich layers and evaporites. During the Tertiary period (between 65 and 2 million years ago), these deposits were folded and pushed upward by the movement of the Arabian continental plate toward the Asian plate. Anticlines in these sediments became traps for the oil that was produced from the decomposition of organic matter in the sedimentary sequence over tens of thousands of years. Such oil reservoirs are found in the Jurassic limestones and dolomites of Saudi Arabia, Cretaceous sands and limestones to the northeast, and in Tertiary limestones and

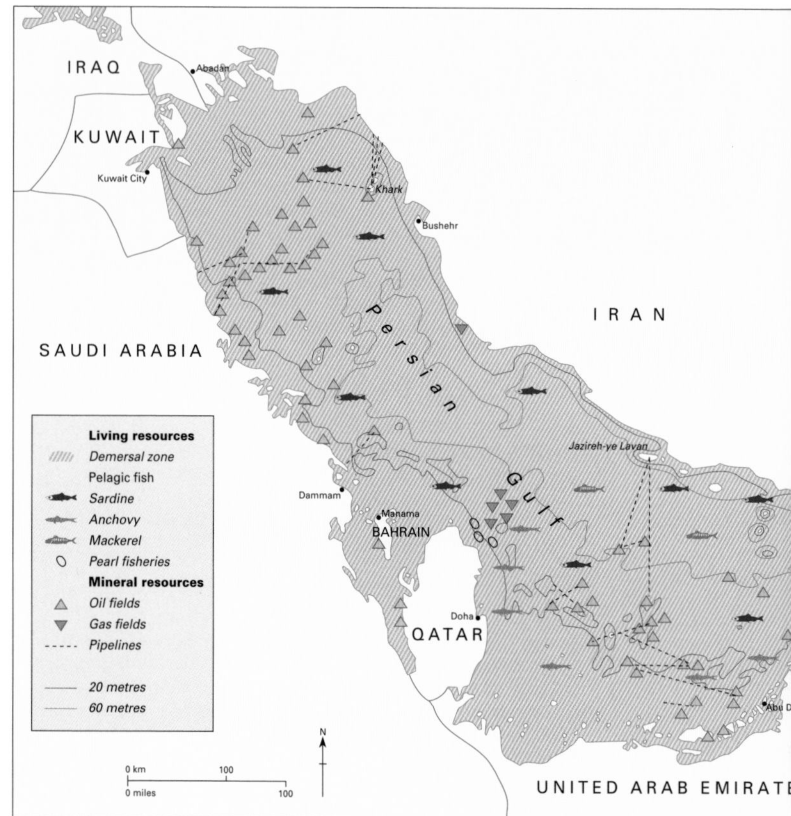

▲ ***Map of the Persian Gulf*** *and locations of the most valuable resources in the region. The area is particularly rich in natural oil.*

sometimes reef structures in the Iranian foothills. A comparatively small field in Triassic rocks lies in Iraq. Gas fields are located in a number of areas in the foothills of the Zagros Mountains of Iran.

Oil from both the inland and offshore fields is pumped through pipelines, which in the case of Das Island in Abu Dhabi is 55 miles (90 kilometers) long, to land-based terminals. Once at the terminal, the oil and gas are separated, and the crude oil refined or pumped to tankers at offshore loading berths.

Following the hostilities in the Gulf during the 1991 Gulf War, around 5 million barrels of oil were released into the sea. Apart from threatening the coastal infrastructure, including desalination plants, this action not only caused the deaths of many seabirds and other marine animals, but also resulted in the degradation of reef areas where the heavier oil fractions covered fragile reef surfaces. Fortunately, the recovery of reefs in the Gulf was more rapid than had been anticipated.

► ***Arab fishing boats*** *off the coast of Qatar. Fishing is a major source of food in the Persian Gulf and the coastal coral reefs are an important marine resource.*

Living resources

Historical references to the rich pearl oyster communities of Bahrain date back to the Assyrians more than 2000 years ago. The invasion of the area by the Portuguese in 1522, was, in part, a desire to gain control over this resource. Exploitation continued until the 1930s through collection of oysters by free divers who also collected freshwater in leather bottles from submarine springs.

At Masirah Island, 9 miles (15 kilometers) off the coast of Oman, there is an important fishery for green turtle, with more than 1000 annually harvested in the seagrass meadows around the island. The island is also an important rookery for loggerhead and other turtle species, with an estimated 30,000 animals nesting there each year.

Commercial fisheries in the Persian Gulf are based on pelagic species, including sardine, anchovy, mackerel and barracuda, and some trawl fisheries are based on demersal species. In general, smaller-scale artisanal fisheries target a much wider range of demersal species.

Trading networks

The prosperous trading civilization of Dilmun encompassed what is now Bahrain and Saudi Arabia, and flourished some 4000–5000 years ago. Taking advantage of the seasonal reversal of winds, mariners could traverse the entire northern Indian Ocean basin, passing down the Arabian and East African coast in search of wood and charcoal. Arab traders carried valuable cowry shells from the Maldives to northern India. They also traveled round the tip of India into southeast Asia, where they established a trading empire dealing in Chinese porcelain, spices, grains, dried fish, ointments and slaves. This network of sultanates and trading centers reached as far as New Guinea to the southeast and the Chinese mainland in the north, and predated the appearance of Europeans in this area.

The Pacific Ocean
Area: *64,000,000 sq miles (166,000,000 sq km)*
Volume: *173,625,000 cu miles (723,700,000 cu km)*
Average depth: *14,050 ft (4820 m)*
Maximum depth: *(Mariana Trench) 36,161 ft (11,022 m)*

The Pacific Ocean covers over a third of the surface of th globe and contains some 174 million cubic miles (72 million cubic kilometers) of water. In area, the Pacific is twic the size of the Atlantic and contains more than twice a much water. Although the Pacific Ocean is the world's larges ocean, it has been decreasing in size throughout its histor as a consequence of the opening of the Atlantic and India Oceans. The margins of the Pacific Ocean are, therefore active areas of subduction, with associated deep-ocea trenches around its perimeter and intense volcanic activity i the region known as the Pacific 'Ring of Fire'.

The eastern half of the ocean basin is characterized by comparatively smooth bottom topography sloping gentl away from the North American coastline. In contrast, the western half has a rugged surface with numerous trenches, both active and inactive, and volcanic island arcs, crumpled and deformed by the northward movement of the Australian continental plate.

The Mariana Trench, where the Pacific plate plunges beneath the Philippine plate at a rate of more than 4 inches (11 centimeters) a year, and the Tonga Trench, north of New Zealand, reach depths of more than 6.5 miles (10.5 kilometers), more than twice the average depth of the ocean. In addition to

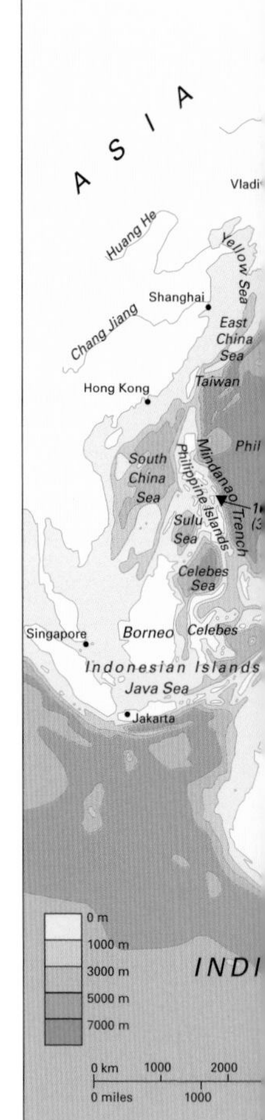

▼ ***False-color*** *image of the Pacific Ocean. The Pacific 'Ring of Fire' is visible at the top of the ocean.*

having a rapid rate of subduction at the plate margins, the Pacific plate also has one of the most active midocean ridges, the East Pacific Rise. On this midocean ridge, new oceanic crust is being formed at a rate of approximately 6.5 inches (15 centimeters) per year.

Eastern Pacific

The floor of the eastern half of the Pacific Basin is relatively simple in structure. Its topography is dominated by the East Pacific Rise and two less active spreading ridges: the Galapagos, which has a spreading center near the equator, and the Chile Rise, which lies southeast of the basin. The slope of the Eastern Pacific seafloor is generally away from the crest of these oceanic ridges.

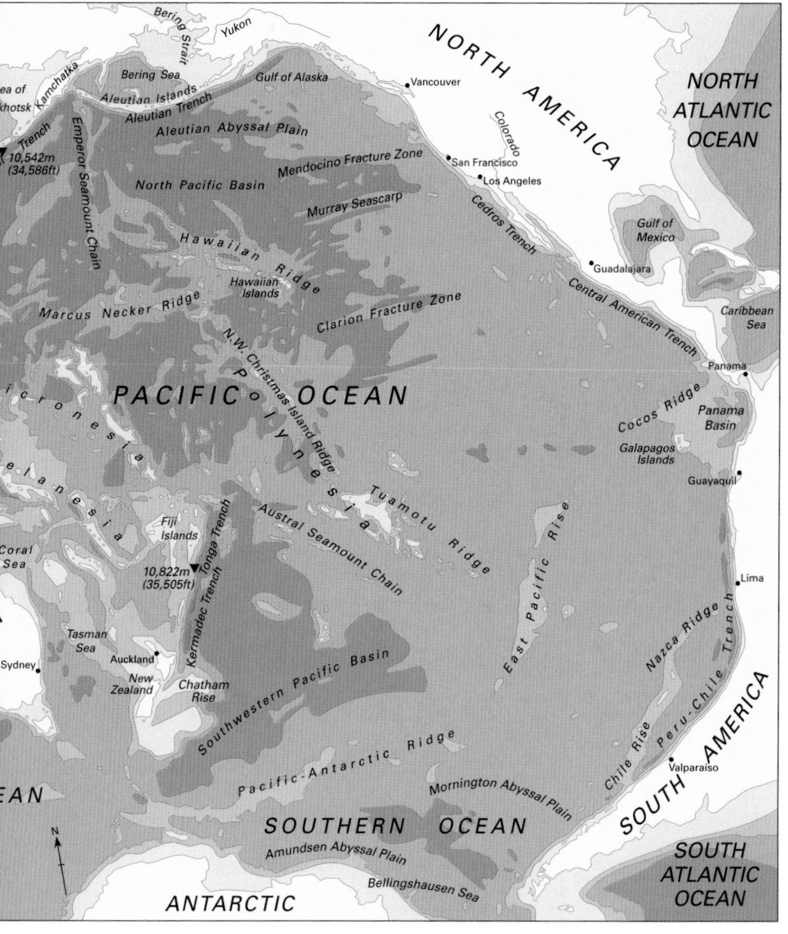

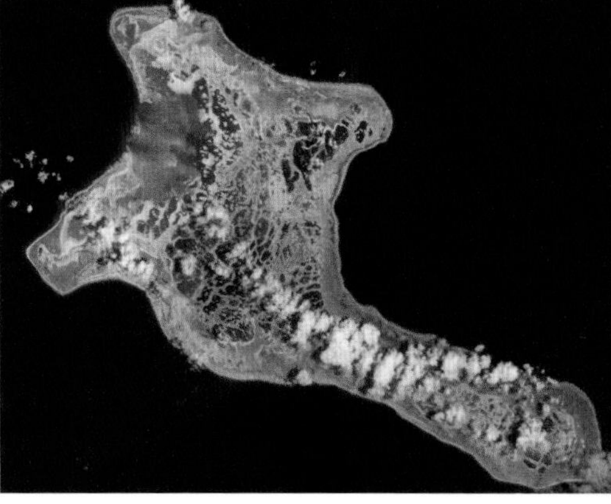

Along the western coast of North America, the sea floor deepens from east to west and increases in age as one passes from Baja California to Hawaii. Baja California is moving away from the rest of North America, and this movement results from a branch of the East Pacific Rise. In the rest of the northeastern Pacific, this midocean ridge has, with the exception of a few remnants off the coast of Washington state, been consumed beneath the continental margin. Thus the midocean ridge, which created the floor of the northeastern Pacific, is no longer visible, having disappeared beneath the North American land mass within the last 30 million years.

Western Pacific

The floor of the western half of the Pacific has a more complex structure. In part, this reflects the separation of the Australian plate from the Antarctic land mass some 55 million years ago, and the shift in the direction of movement of the Pacific plate. The Australian plate is continuing to move northward past the East Indies, which further compresses the oceanic crust at its northern margin. Paired active trenches, separated by a number of both active and inactive trenches, also mark this western region of the ocean floor. The Philippine Sea, for example, has been created in three distinct phases, each marked by the formation of a separate basin between the trenches.

The oldest oceanic crust is located in the western Pacific, where some areas of the ocean floor are between 100 and 135 million years old. Its structure has been complicated by subsequent volcanic activity, and chains of volcanoes such as those forming the Gilbert and Ellis Islands, and the Emperor seamounts, are widespread. The western Pacific also has several large volcanic plateaus, including the Shatsky Rise, the Solomon Plateau and the Manihiki Plateau. These have apparently been formed by extensive lava flows similar to the ones that created continental features such as the Deccan Plateau in India.

Pacific islands

The Pacific Ocean Basin is characterized by numerous island chains running in a general northwest-southeast direction. The oldest of these island chains show a greater tendency toward a north-south orientation than those which are less than 40 million years old. This difference in orientation indi-

Water profiles of the Pacific Ocean

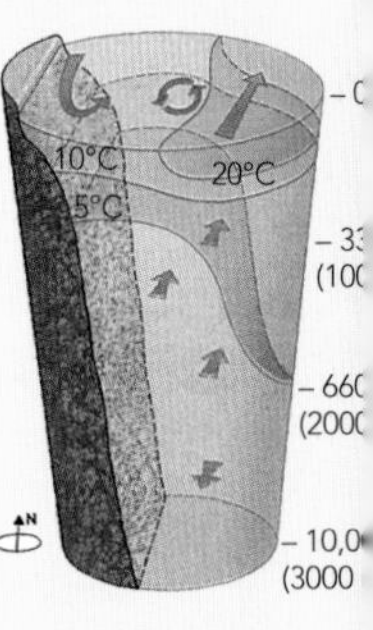

1 Kuroshio Current

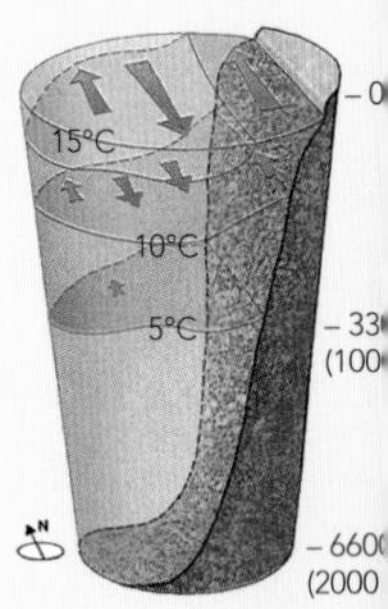

3 California Current

◀ ***Kiritimati** is one of the Northern Line Islands in the western Pacific Ocean. It is the world's largest atoll (125 square miles/321 square km). The lagoon is nearly completely filled by coral growth. Captain James Cook named the atoll Christmas Island when he arrived on Christmas Eve, 1777. Used for nuclear tests in the 1950s and 1960s, Kiritimati is valued for its marine and wildlife resources – 6 million birds breed here.*

cates a shift in the direction of movement of the Pacific Ocean crust around 40 million years ago.

Atoll islands are characteristic of the Pacific basin and are composed of small piles of sand on top of living corals that are themselves growing on the skeletons of previous generations of coral. Charles Darwin was the first person to propose that such islands were formed on submerged seamounts that were slowly sinking. The upward growth of corals keeps pace with the slowly sinking land beneath. Nearly 100 years later, in 1952, evidence in support of this theory came from Enewetak Atoll where deep drilling revealed that 4300 feet (1300 meters) of carbonate limestone overlay the summit of an ancient volcanic cone. Throughout the region, one can see islands in different stages of atoll formation, from the recently emerged rugged volcanic islands with their small fringing reefs to islands where the volcanic cones are eroded and the reefs found further offshore protect a sheltered lagoon between the reef margin and the land.

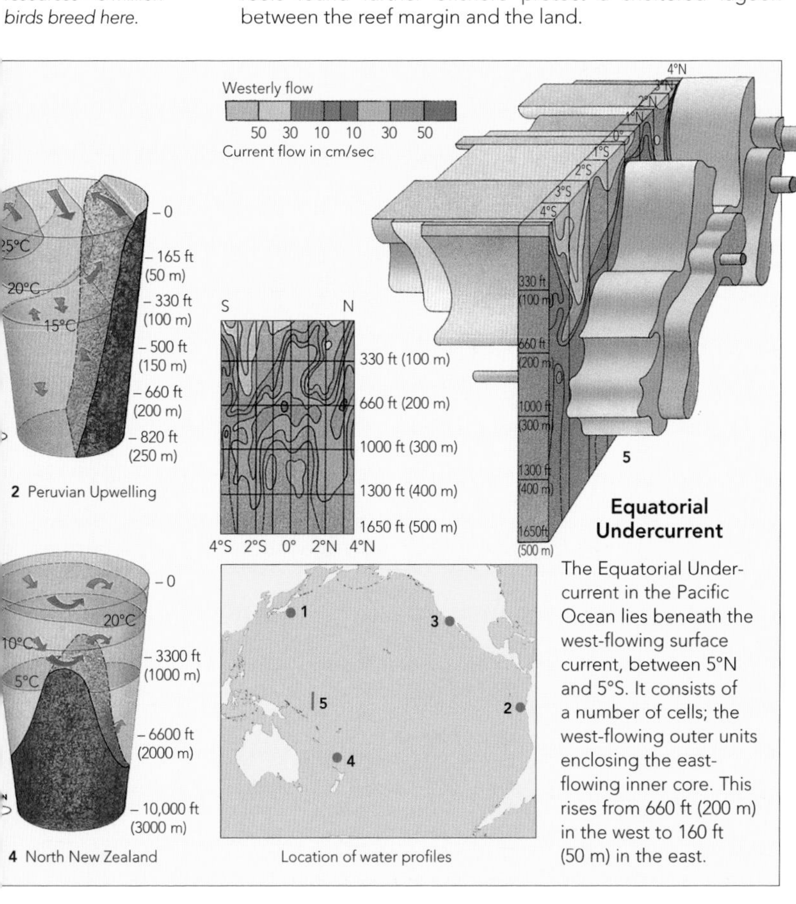

2 Peruvian Upwelling

4 North New Zealand

Location of water profiles

Equatorial Undercurrent

The Equatorial Undercurrent in the Pacific Ocean lies beneath the west-flowing surface current, between 5°N and 5°S. It consists of a number of cells; the west-flowing outer units enclosing the east-flowing inner core. This rises from 660 ft (200 m) in the west to 160 ft (50 m) in the east.

▼ ***Younger chains** of islands run northwest-southeast, in contrast to those older than 40 million years, which run north-south.*

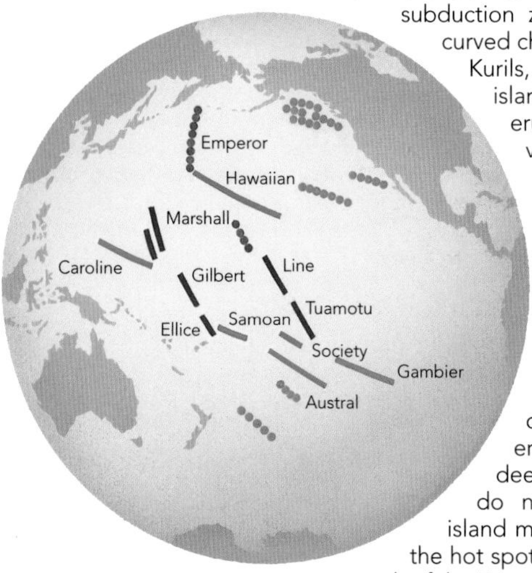

Less than 40 million years old

islands

seamounts

Over 40 million years old

islands

seamounts

In addition to the true oceanic islands, which have never been part of a larger land mass, islands like New Zealand represent small fragments of continental crust which have become separated from the larger continental land masses. The oceanic islands are all volcanic, and in the Pacific, they are of two distinct types. Along the landward side of subduction zones in the western Pacific are curved chains or arcs of islands, such as the Kurils, Bonins and Marianas. These islands were formed by the explosive eruptions of andesite lavas from the volcanoes on top of the subduction zones, where oceanic crust is being consumed beneath the continental plate margin.

In contrast, the straighter island chains of the mid-Pacific are formed of basaltic lavas, which erupt much less violently. These islands are apparently formed over hot spots in the oceanic crust, where intermittent eruption of lava from a source in the deep mantle occurs. These hot spots do not erupt continuously, and the island may move with the crust away from the hot spot. Thus the oldest islands lie to the north of the Hawaiian chain.

Beyond the emergent islands, the chains may continue beneath the sea surface as submerged seamounts, or guyots. The Emperor seamount chain (running north-south) represents a continuation of the younger Hawaiian island chain.

Current circulation

The two main gyres of the Pacific Basin are separated by a more complex system of equatorial currents and counter-currents than the Atlantic. In addition, anomalous patterns of circulation can occur during El Niño years. During such events, the surface currents moving westward weaken and warmer water from the western Pacific flows back toward the Latin American coast, shutting off the Peruvian upwelling.

In the northern hemisphere, the warm North Equatorial Current carries water some 9000 miles (14,500 kilometers) across the Pacific, the longest westward-flowing current in the world's oceans. At the western end, the current turns north and is intensified into the narrow western boundary current, the Kuroshio, which flows northward at speeds in excess of 90 miles (145 kilometers) a day. In the latitude of Japan, the Kuroshio meets the cold, south-flowing Oyashio Current, and the two are deflected away from the coast to meander eastward across the North Pacific. To the north of this warm-water gyre, lies a cold, subpolar gyre formed by the Alaskan and Aleutian currents in the east, and the Oyashio Current in the west.

South of the equator, a second major gyre is formed by the convergence of the South Equatorial Current, the East Australia Current and the Humboldt Current. The Humboldt Current is stronger than most eastern-boundary currents, due to the northward deflection of the westerly winds as they meet the Andes. The western boundary current, called the East Australia Current, turns westward at approximately the latitude of Sydney and passes to the north of the North Island of New Zealand. The Antarctic Circumpolar Current is driven by the westerly winds and has a surface speed of only 12 miles (19 kilometers) per day. The total volume of water carried amounts to more than 165 million tons per day, greater than the volume transported by any other current in the world's oceans. This current reaches down to depths of 2 miles (3 kilometers), hence the immense volumes of water transported by the current.

Countercurrents

In the region of the equator, the flow is complicated by the presence of the North and South Equatorial countercurrents, which result in the Equatorial Undercurrent. This flows east at speeds of more than 90 miles (145 kilometers) per day. The main body of the current lies around 660 feet (200 meters) below the surface in the west, rising to around 160 feet (50 meters) in the east.

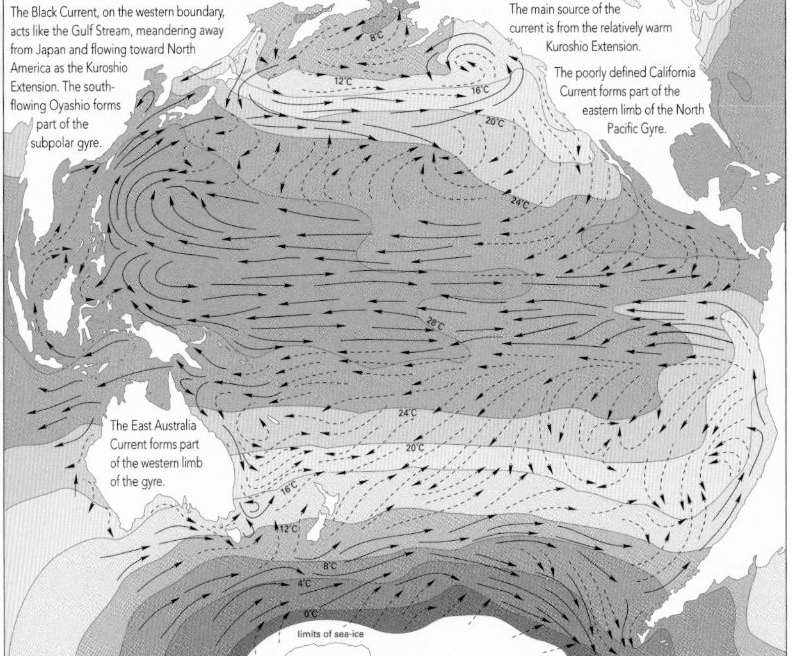

▼ ***Map of currents*** *in the Pacific Ocean. Surface temperature varies enormously in the region, ranging from 32°F (8°C) in the Bering Strait, to 84°F (29°C) off Micronesia.*

PACIFIC OCEAN RESOURCES

For many isolated, insular nations of the central Pacific, living marine resources are the sole source of export income for development. The potential of deep-sea mineral resources has yet to yield economic benefit.

Nonliving resources

Polymetallic nodules are widely distributed throughout the deep Pacific. By 1974, 100 years after their initial discovery, it had been established that a broad belt of seafloor – across an area of 1.35 million square miles (2.15 million square kilometers) – between Hawaii and Mexico was densely paved with such nodules. Despite their relative abundance, mining the nodules remains uneconomic. Similarly, deposits of phosphate minerals in extensive areas off the western coasts of North America and South America, and on submarine plateaus regions near New Zealand and Australia, cannot be economically extracted at present.

Near shore, many mineral placer deposits are found on the continental shelves surrounding the Pacific. Gold has been mined from the beaches of Alaska for many years, and platinum is known to occur in the same region. Tin is mined extensively in the southeast Asian region, and mineral sands containing titanium, chromium and zirconium are present along the North American continental margin. In the western Pacific, extensive offshore deposits of iron ore have been mined by the Japanese for many years.

Living resources

In excess of 40 percent of the world's harvest of finfish comes from the Pacific, and the bulk of this is composed of herring-

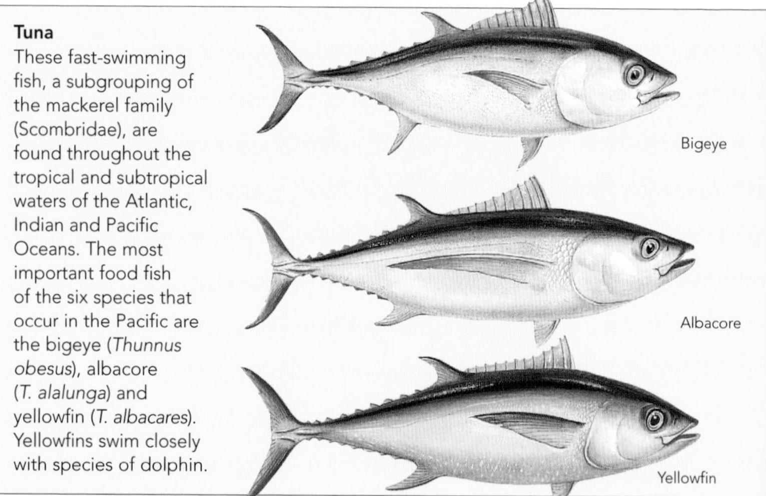

Tuna

These fast-swimming fish, a subgrouping of the mackerel family (Scombridae), are found throughout the tropical and subtropical waters of the Atlantic, Indian and Pacific Oceans. The most important food fish of the six species that occur in the Pacific are the bigeye (*Thunnus obesus*), albacore (*T. alalunga*) and yellowfin (*T. albacares*). Yellowfins swim closely with species of dolphin.

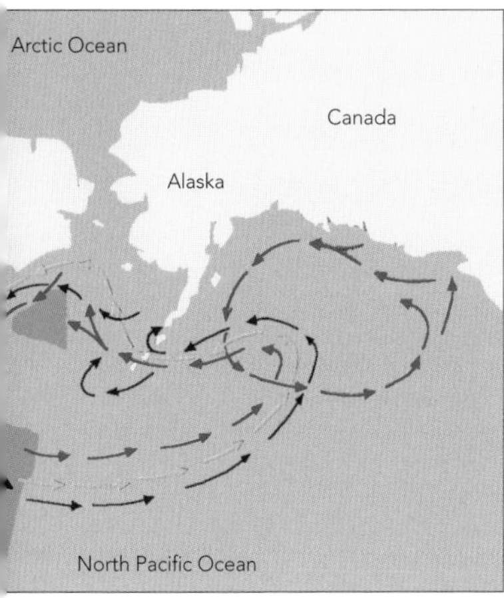

▲ ***The six species*** *of Pacific salmon migrate from their spawning grounds in the rivers of Asia and North America to the rich feeding grounds of the North Pacific. After periods ranging from six months to five years at sea, the adults return to the rivers where they were born to spawn, making journeys of thousands of kilometers.*

like fishes, including the Japanese sardine and Peruvian anchovy. Between 1969 and 1971, the Peruvian anchovy fishery had risen to almost 30 percent by weight of the total fishery of the Pacific, one-sixth of the world catch of fish. The harvest of anchovy, which was almost exclusively for the manufacture of fishmeal, exceeded 10 million tons in the peak years. But by 1991, the catch dropped by more than half to around 4 million tons. The cause of this was the El Niño events of the previous decade, but the cause of fluctuations in other Pacific fisheries, such as the huge growth during the 1930s and 1970s of Japanese sardine stocks, are not well understood.

Six species of Pacific salmon migrate from spawning grounds in the rivers of Asia and North America to the rich feeding grounds of the North Pacific, remaining at sea for periods of six months to five years before returning to freshwater to spawn. Like the albacore tuna, which travels more than 5300 miles (8500 kilometers) from California to Japan, these species are fished by both the Japanese and the Americans.

The Pacific tuna fishery is based on purse-seine fisheries in the eastern Pacific and long-liners in the west and central Pacific regions. These commercial fisheries target the deeper-swimming species of albacore, bigeye and yellowfin tuna, and the catch is frequently processed at sea. The recent introduction of extensive drift nets, several tens of kilometers in length, is a cause of concern to environmental groups and the smaller nations of the Pacific for whom tuna represent the sole, exportable resource. These drift nets, used in the Pacific by long-distance fishing fleets from the East Asian region, have been termed the 'wall of death', since they result in the entanglement and drowning of turtles, marine mammals and a wide variety of nonresource species. A regional convention has been drafted to ban their use in the South Pacific.

Another high-value species is the North Pacific salmon, taken generally inshore using gill nets, traps and weirs – although they are also fished on the high seas by Japanese fishermen using gill nets and purse seines. Other demersal fisheries include the Alaskan pollack fishery in the subarctic Pacific.

Artisanal fisheries

In the island nations of the central Pacific, the most important fisheries are small-scale artisanal fisheries based on the wide diversity of species associated with the coral-reef ecosystems which surround these islands. The enormous diversity of species in these catches presents problems of processing and marketing outside these countries, where the unfamiliar characteristics of many of the species result in low international market acceptability.

The Antarctic Ocean
Area: *13,513,000 sq miles (35,000,000 sq km)*
Sea ice: *8,100,000 sq miles (21,000,000 sq km) freeze in winter: 1,540,000 sq miles (4,000,000 sq km) permanently frozen*

The Antarctic, or Southern Ocean, is normally taken as meaning the area of ocean which lies south of the Antarctic convergence zone, or between 50° and 55°S, and covers an area of approximately 35 million square kilometers (13.5 million square miles). The Antarctic convergence zone represents the invisible boundary where the cold surface waters of the Antarctic Ocean meet the warmer waters of the sub-Antarctic.

The Southern Ocean forms a great expanse of unbroken ocean. This vast area of water is important in the formation of cold water which flows at the surface and at depth into the Atlantic, Pacific and Indian Ocean basins, and is replaced by southward-flowing intermediate, warmer water from sub-Antarctic areas.

The Antarctic icecap

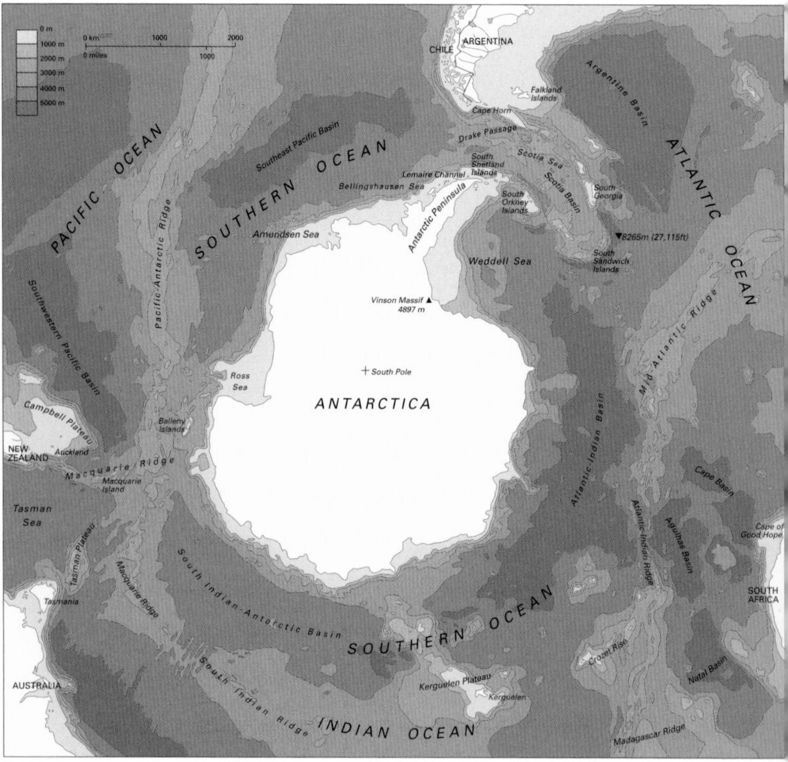

Antarctica is the coldest continent, with temperatures in the central region rising briefly to 86°F (30°C) in the two warmest months, and sinking to –85°F (–65°C) in winter. A narrow coastal fringe separates the icecap from the ocean. On this area of land, which represents less than 2 percent of the surface of the continent, the majority of the Antarctic wildlife lives and breeds. This is also the area in which more than 70 permanently manned research stations have been established by more than 20 nations. Seven nations have territorial claims to this continent, claims which were recognized but suspended under the 1959 Antarctic Treaty.

Almost the entire land mass of Antarctica lies within the Antarctic Circle, at the center of which lies the South Pole. Two deep indentations disrupt the circular outline of this continent: the Weddell Sea, which faces the South Atlantic, and the Ross Sea, which faces out toward the South Pacific Ocean.

During the Antarctic summer, this inhospitable continent is fringed with 4 million square kilometers (1.5 million square miles) of ice-covered sea. During winter, the ice extends to cover an area of some 8 million square miles (21 million square kilometers). The continent itself supports an icecap which, over time, has accumulated to an average depth of 6500 feet (2000 meters) and which extends to more than 10,000 feet (3000 meters) in some areas.

▲ ***Emperor penguins*** *are the largest of all the penguins. After mating, the female lays one egg, and returns to the sea, leaving the incubation to the male.*

The Antarctic ice sheet contains more than 90 percent of the world's ice and around 70 percent of the planet's freshwater. Glaciers move seaward at numerous points to feed the floating ice shelves which border the continent and extend out several hundred kilometers to a submerged depth of between 500 and 650 feet (150 and 200 meters). Ice shelves cover the head of the Ross Sea and the inland and western shores of the Weddell Sea.

Depth measurements

The continental shelf of Antarctica is generally much narrower and deeper than that of other continents. It lies at a depth of between 1200 and 1600 feet (370 and 490 meters), and there is a deep depression between the outer edge of the continental shelf and the land, believed to be the result of sinking of the central Antarctic continental land mass under the weight of the overlying ice sheet.

Beyond the continental shelf lies the deep-ocean basin, which is bounded to the north by the midocean ridge systems. The ocean basin reaches depths of between 13,000 and 16,500 feet (4000 and 5000 meters) and is subdivided into the southeast Pacific, southern Indian and Atlantic-Indian basins by ridges which join Antarctica to America, Antarctica to the Kerguelen Plateau, east of the South Indian Basin, and Antarctica to Tasmania.

Ocean circulation

The surface currents of the Antarctic Ocean are driven by the Roaring Forties, the Furious Fifties and the Shrieking Sixties, and its southern boundary is the ice and rocks of the Antarctic coastline. North of 60°S, the current flow of the Southern Ocean at all depths is to the east. South of 60°S, the flow is to the west, but both current systems contribute to the northward flow of cold water in the surface and deeper layers of the ocean.

In the northern half of the Antarctic Ocean, the southerly movement of water is generally below 6000 feet (1800 meters), but around the Antarctic Convergence Zone, it rises sharply to within 300 feet (100 meters) or so of the surface. The warmer subantarctic waters overlie a column of cold water which occurs at depths of between 6500 and 10,000 feet (1900 and 3000 meters). This sharp transition zone is known as the polar front and represents the point at which the cooler, dense Antarctic surface water sinks beneath the warmer subAntarctic surface waters. At the surface, this sub-Antarctic convergence zone is recognizable by a sharp rise in temperature of between 1° and 3°F (1° and 2°C) and a change in the composition of the plankton community.

Some 10° further north from the polar front is found the less well-defined zone of convergence between the subantarctic and subtropical waters. Although not well-defined as a current boundary, it is recognizable at the surface as a transition zone between the colder waters of southern origin and the warmer, more saline water characteristic of regions of higher latitude.

Life in the frozen seas

Despite the icy seas and frozen wastes of the Antarctic, the Southern Ocean is home to a wide variety of seabirds, seals and whales. These marine animals depend on the biological productivity of the microscopic phytoplankton and the kelp, which fringe the shores of the continent. The long daylight hours of the Antarctic summer result in high productivity of

◀ ***Tail of a humpback whale*** *Humpbacks are well known for their acrobatic behavior. They were once plentiful in the Antarctic, but intensive whaling in the first half of the 20th century has brought numbers down to the low thousands.*

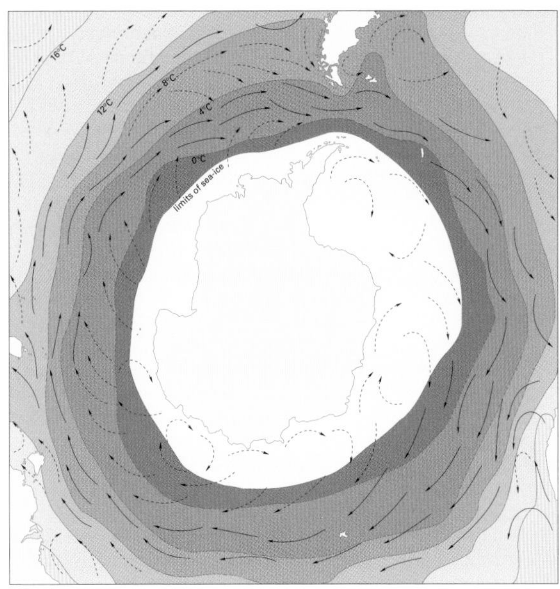

► ***Map of the Southern Ocean*** *showing circulation patterns and sea-surface temperature. There is a significant change in temperature at the subantarctic convergence zone.*

phytoplankton estimated at 610 million tons a year. This vast community of phytoplankton is, in turn, eaten by zooplankton of which half are krill that form the major food source for fish, squid, whales and some seabirds.

Of great concern at the present time is the hole in the ozone layer, which may have impacts on the phytoplankton and kelp in the region. Depletion of the atmospheric ozone layer allows higher levels of ultraviolet radiation to penetrate to the surface of the ocean, which affects plankton productivity and could ultimately affect the entire Antarctic ecosystem.

Seven species of filter-feeding whales and eight species of toothed whales inhabit this ocean, although none is confined to this area. The baleen whales, which feed predominantly on krill, undertake migrations to warmer, equatorial waters during the southern winter, returning to the Antarctic during the summer. The substantial reduction in numbers of filter-feeding baleen whales must have had a huge impact on the populations of krill – the plankton on which they feed. It has been estimated that before they were exploited, the southern baleen whales alone were consuming around 190 million tons of krill each year.

Although most seabirds are found in the sub-Antarctic region, a number, including three species of penguin, breed on the Antarctic land mass itself. The breeding colonies of seabirds are densely packed since the extent of ice-free land is limited to the narrow coastal fringe. Antarctic seals, of which there are six species, are also abundant, including an estimated population of around 15 million individual crab-eater seals.

ANTARCTIC RESOURCES

The history of resource use in Antarctica is one of short-term perspectives dominated by greed and competition. For example, during the 1930–31 whaling season, nearly 43,000 whales were killed, and the present whale populations are probably one-sixth to one-tenth of their numbers prior to whaling. As individual species of whales became depleted, the industry shifted to more profitable species. For example, between 1927 and 1936, the humpback whale had become unprofitable and was replaced in the industry by the blue whale. In postwar years, the blue whale was replaced by the fin whale in turn, which was supplanted by the smaller sei, bottlenose and minke whales.

It has been estimated that there are at present only about 500 blue whales in the Antarctic, although they may have once numbered 250,000. The number of sei whales is more uncertain, due to the fact that their entire range has not been surveyed, but they may have been reduced to a comparable extent. Humpback and right whales, which originally numbered around 100,000, have been reduced to the low thousands. The present status of whale stocks is very uncertain due to the difficulties of surveying these animals, and it will be several decades before firm evidence for a recovery of the most heavily depleted stock becomes available.

This dramatic decline in numbers of whale species results, in part, from the absence until recently of any international agreement concerning the exploitation of Antarctic resources. The situation has improved following the negotiation of a variety of international treaties, including in 1978 the 'Convention for the Conservation of Antarctic Seals' and in 1982 the more far-reaching 'Convention on the Conservation of Antarctic Marine Living Resources'.

▼ ***A research vessel*** *approaches the Antarctic coast along the Lemaire Channel. Antarctica's main human population is scientists from all around the world. Research ranges from counting whale numbers to the search for mineral resources.*

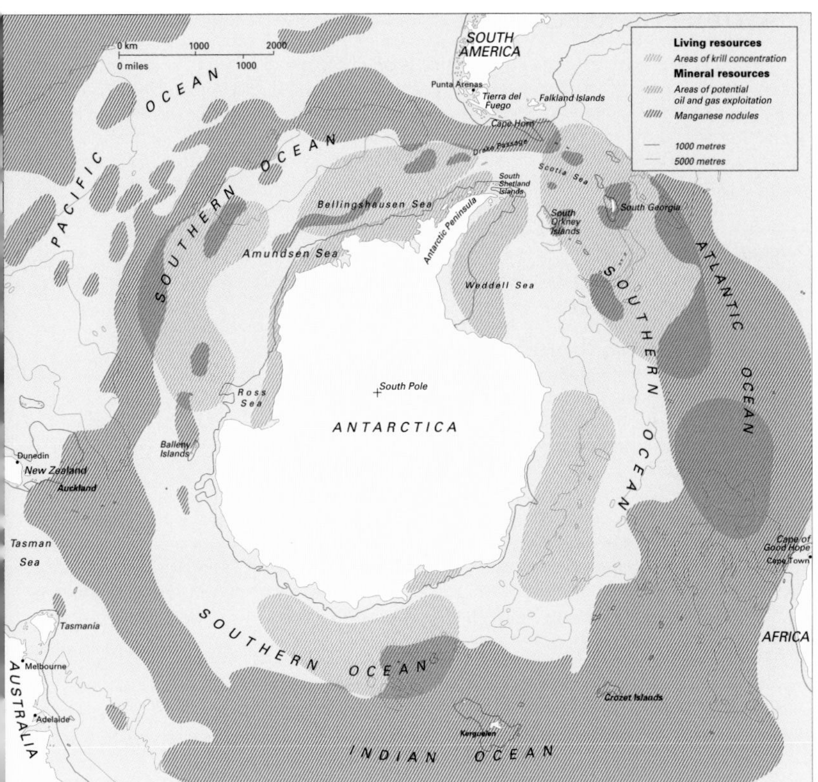

▲ ***Map of the Southern Ocean*** *showing the location of natural resources. Despite there being concentrations of mineral resources, the expense of extraction is currently greater than the need. Krill is harvested to produce animal feed.*

Less well known than the overexploitation of the whales is the history of the Antarctic fur seal. This particular species of seal was practically exterminated by the 1820s after several hundred thousand had been slaughtered for their skins. The seals subsequently recovered to some extent, but were again overexploited in the 1870s, and are now recovering once more.

Nonliving resources

Iron and coal deposits have been found in the mountain ranges of the Antarctic continent, while the deep-drilling ship, *Glomar Challenger*, found deposits of natural gas in the sediments beneath the Ross Sea. Since Antarctica was once part of the larger land mass of Gondwanaland, it is expected that mineral reserves comparable to those of other southern continents are likely to occur within the rocks of Antarctica. At present, the economic costs of mineral extraction preclude their exploitation. The inhospitable climate, together with the thickness of the icecap and the problems of exporting the ores once mined, remain the greatest safeguards against unrestrained mineral exploitation.

PROKARYOTES AND EUKARYOTES

All living organisms fall into two major groups: the eukaryotes, which possess chromosomes contained in a defined nucleus, and the prokaryotes, such as bacteria, which lack a nucleus and chromosomes.

CYANOBACTERIA AND BACTERIA

The cyanobacteria (blue-green algae) photosynthesize. Not all are blue-green, some are yellow, red, purple or black, depending on the types of photosynthetic pigments they contain. Most marine blue-green algae are found in shallow waters intertidally or subtidally, living attached to other plants or rocks. Blue-green algae can reproduce at a remarkably fast rate, forming blooms within a few hours and imparting color to the water. *Oscillatoria erythrea*, a red-colored species, gives the Red Sea its name.

Marine bacteria are most abundant around the coasts, but in the open ocean they occur in association with plankton. Many are saprophytes, which degrade organic materials, including chitin, celluloses, mucins, and fats. Decomposing bacteria are as important in the marine environment as on land, while some chemotrophic species form the basis for the food chains of submarine thermal vents and others, the sulfur-reducing bacteria, are important in the anaerobic conditions of intertidal muds.

KINGDOM FUNGI

Molds, yeasts and mushrooms form one of the five kingdoms of living organisms. They are heterotrophic, lack photosynthetic pigments and may dominate the microbial community in coastal waters. In the marine environment, the dominant fungi are the saprophytic yeasts, which feed on dead organic matter.

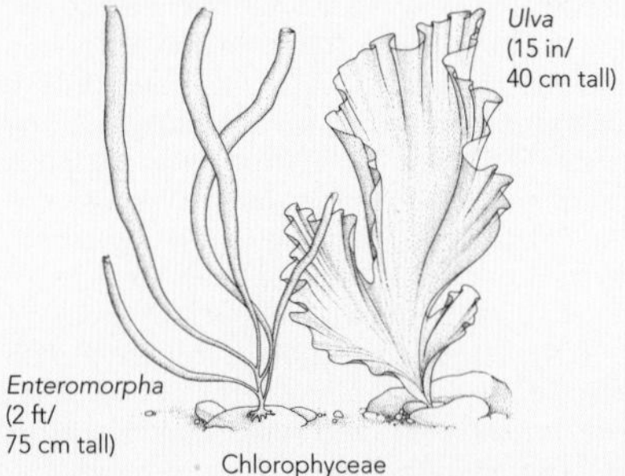

Chlorophyceae

THE PLANT KINGDOM

Plants form the basis of all life both on land and in the sea. By converting sunlight to chemical energy through the process of photosynthesis (autotrophy or self feeding), plants make energy available to heterotrophic organisms. The dependence of autotrophic plants on sunlight limits the depth range of these organisms. In shallow coastal waters, single-celled and multicelled algae may grow attached to the surface of the substrate, while in deeper waters, small planktonic, single-celled organisms form the basis of the food chain.

SUPER-PHYLUM ALGAE

The algae range from microscopic, unicellular organisms that reproduce by simple division, to those that are extremely large, such as the giant kelps, which can attain lengths of up to 165 feet (50 meters). The seaweeds have complex life histories and reproductive cycles, and different tissues, which may include floats, holdfasts for attachment to the substrate, and different male and female organs. Seaweeds are classified according to their pigments, and while all contain the typical green pigment chlorophyll-a, various yellow, brown or red pigments may mask the green coloration completely.

Class Phaeophyceae

The brown seaweeds contain chlorophyll-c, are colored brown by fucoxanthin pigments and store the carbohydrate laminarin as a food reserve. They range in form from simple filamentous types to the large kelps, *Laminaria* spp., which have long bladelike fronds and a strong stalk and well-developed holdfast for attachment, which can grow at the rate of 4 feet (1.25 meters) per day when cut. They are harvested in many areas as sources of mannitol (a sweetening agent), oil and alginates.

The most familiar brown seaweeds are the fucoids, common in the intertidal zone of temperate rocky shores. These species have air-filled bladders which support the fronds in the water, while the plant as a whole is resistant to exposure to the air, high temperatures and drying. Most of these plants live between two and four years, although some species may survive for up to 19 years. *Sargassum*, a brown, floating fucoid, occurs in enormous masses in the Sargasso Sea.

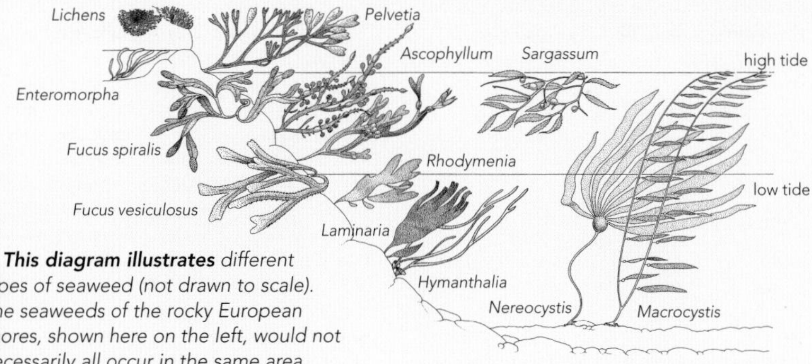

► ***This diagram illustrates** different types of seaweed (not drawn to scale). The seaweeds of the rocky European shores, shown here on the left, would not necessarily all occur in the same area.*

Class Rhodophyceae

The red seaweeds colored by phycocyanin, phycoerythrin and lutein pigments, include species of pink, violet, purple, red and brown coloration. Most species live attached to the substrate in the coastal zone and are flattened or filamentous. Some have a calcareous, encrusting growth form (e.g. *Lithothamnion*), while the flattened *Porphyra* is grown commercially and eaten in Japan. The most important commercial product from red algae is agar or carrageen, the growth medium used extensively in bacterial culture in medicine.

Class Chlorophyceae

Some of the unicellular green algae are motile with flagella, such as *Pyramimonas* and *Tetrahele*, while nonmotile forms are found intertidally, including the large, multicellular genera, *Enteromorpha* and *Ulva*. Some of the larger forms are collected for food, but economically their importance relates more to their fouling of ship hulls. Most species are larger than the red seaweeds and may be up to 3 feet in length, but the planktonic forms are members of the picoplankton (between 0.2 and 2 μm in length).

Class Prasinophycea

These are planktonic green algae, usually with four flagella, which differ from most of the Chlorophycea in having small scales on the flagella and sometimes over the body surface. Some species are colonial and form a colony within a spherical ball of mucus.

Class Bacillariophycea

The unicellular diatoms are extremely important components of the phytoplankton, forming a food source for filter-feeding animals, such as copepods. Diatoms have a rigid siliceous cell wall, formed of two overlapping valves making a sculptured little box with a closely fitting lid. Holes in the lid allow the entry of nutrients in solution and appear to be involved in locomotion in some species. The siliceous shell is much denser than seawater, and hence many species have spines or hairs projecting from the shell to reduce the rate of sinking.

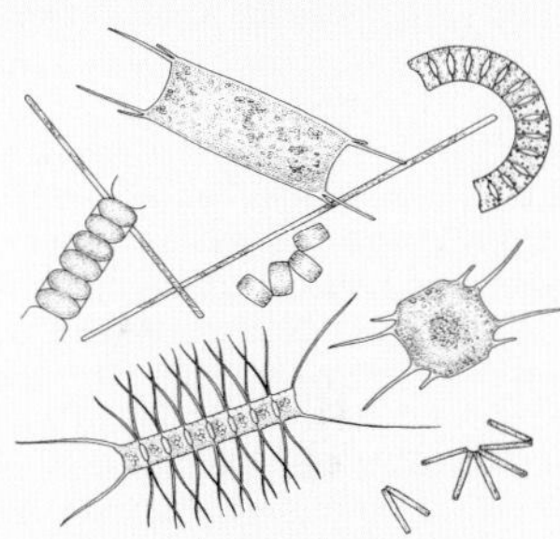

Bacillariophyceae (diatoms)

ANGIOSPERMS

50 species of monocotyledonous plants, the seagrasses, have secondarily invaded the marine environment. They grow in shallow coastal waters and are generally rooted in soft sediments in the intertidal and subtidal zones. Most species produce flowers, and fertilization is by means of floating pollen, although at least one species flowers above water. These plants, like mangroves, are viviparous, the seeds germinating on the parent plant and the young propagule growing to relatively large size before becoming detached and assuming an independent existence.

The center of diversity of seagrasses is the Indowest Pacific with up to 16 species

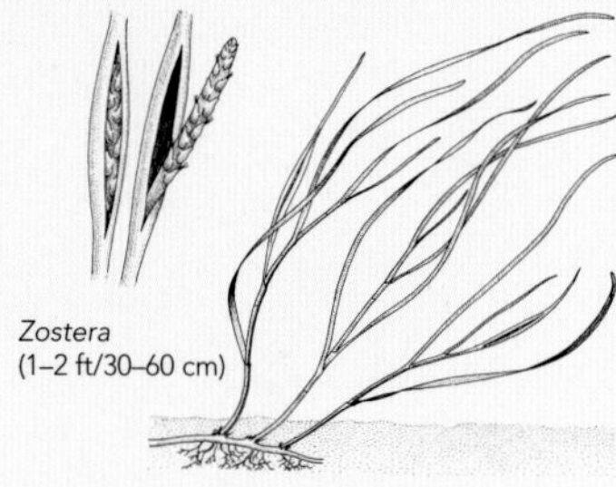
Zostera
(1–2 ft/30–60 cm)

growing in a single location, but seagrasses are also distributed in the Mediterranean, where the single species of *Posidonia* is extremely widespread, and in temperate Europe, where the typical genus, *Zostera*, has declined in recent decades.

KINGDOM PROTISTA

This kingdom includes around 50,000 species of single-celled organisms most of which are motile and heterotrophic. Of the free-living protistans, around two-thirds are marine. Protozoa live in the sea, in freshwater, or as parasites in other organisms. They usually multiply by simple fission (division of one individual into two or more new cells), although sexual reproduction also occurs in many forms. The different phyla are distinguished by their modes of locomotion; the Sarcomastigophora locomote by means of flagella or pseudopodia, the Ciliophora by means of cilia, and the Sporozoa are parasitic.

PHYLUM SARCOMASTIGOPHORA

The dinoflagellates are motile unicellular organisms with two flagella. Some, such as *Noctiluca,* are phosphorescent when disturbed, while others, such as *Gonyaulax,* produce toxic algal blooms. One genus *Symbiodinium* is extremely important as a symbiont of corals and giant clams. The Chrysophyceae, including the so-called silicoflagellates, are mainly found in freshwater but the genera *Dictyota* and *Distephanus* are widespread in marine phytoplankton. The Xanthophyceae have green plastids, one of which, *Vaucheria,* forms mats of greenish threads of several centimeters length in coastal salt marshes, while *Meringosphaera* is often present in tropical marine nanoplankton. A number of species of Mastigophora are colonial, forming large spherical, hollow structures which may contain as many as 10,000 individuals.

The Sarcodina include important marine groups, such as the Amoebida, the Foraminifera, the Heliozoa, the Rhizomastigina and the Radiolaria. Foraminifera construct shells of calcium carbonate which are multichambered, and the majority are bottom-dwelling organisms. Important marine genera, such as *Globigerina*, are planktonic and have long thin spicules which reduce the rate of sinking.

Radiolarians are all planktonic with siliceous skeletons resembling three-dimensional snowflakes. Radiolarian skeletons build up to form the radiolarian ooze of deep tropical basins. The Rhizomastigina are a peculiar group of organisms similar to the Radiolaria, and the remaining group, the Heliozoa, are carnivorous and are capable of subduing copepods and even nematodes.

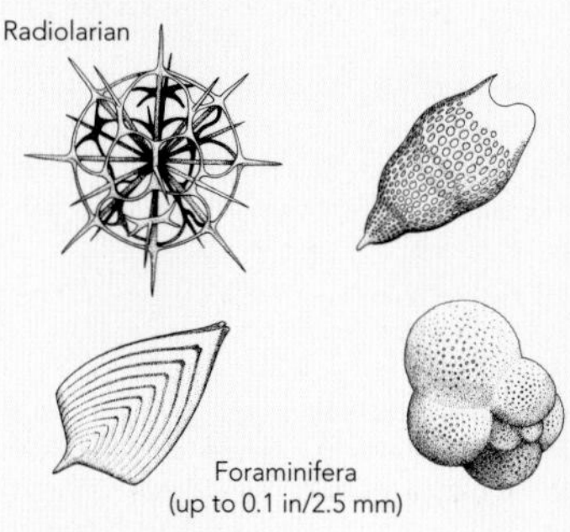

Foraminifera
(up to 0.1 in/2.5 mm)

PHYLUM SPOROZOA

This group is composed of parasitic species many of which are found in the gut of invertebrates or the gall bladder and other tissues of vertebrates, including fish. Sporozoan infections may be common and comparatively harmless to the host under natural conditions, but under the crowded conditions of hatcheries and shellfish beds, they may become a serious economic problem.

PHYLUM CILIOPHORA

The ciliate protozoans number more than 7000 species which move and feed by means of cilia. Marine ciliates occur in the plankton, interstitially and as benthic forms. Simple ciliates are covered with rows of cilia arranged over the surface and around the mouth, while the Tintinnids, a group of around 900 species, have lost the cilia and developed a crown of membranelles around the mouth.

THE ANIMAL KINGDOM

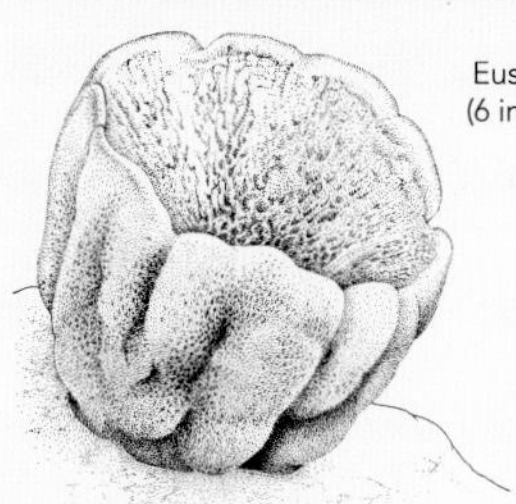
Euspongia (6 in/15 cm)

In terms of the absolute number of species, marine animals are less numerous than on land. There are many phyla of marine organisms which have no terrestrial or freshwater representatives, and hence the diversity of the marine fauna, in terms of higher taxa or genetic diversity, is greater than on land.

PHYLUM PORIFERA

The sponges are sessile animals with a simple body form. They filter-feed by means of cilia, which generate a current of water that flows into the animal through numerous pores over its surface. Food particles are filtered out of this flow of water, which passes out of the animal via one or more excurrent openings. The sponges show various grades of organization, from a cup-shaped structure which lacks folds in the body wall, through sponges where the wall is folded, to the largest and most complex sponges where the central chamber of the sponge is completely filled with tissue. Four classes of sponges are recognized, based in part on the nature of the skeleton: the Hexactinellida or glass sponges are the most primitive of living sponges, with a skeleton of siliceous spicules, fused to form a delicate basketlike structure in species, such as the Venus flower basket; the Calcarea are small, mainly littoral sponges which encrust rocks and are built on a skeleton of two-, three- or four-rayed calcareous spicules; the Desmospongia, the largest class of living sponges, characterized by a complex canal system and a skeleton of either siliceous spicules of spongin fibers or a mixture of both; and the Sclerospongia, a few species that, in addition to an internal skeleton similar to that of the desmospongia, have an outer casing of calcium carbonate.

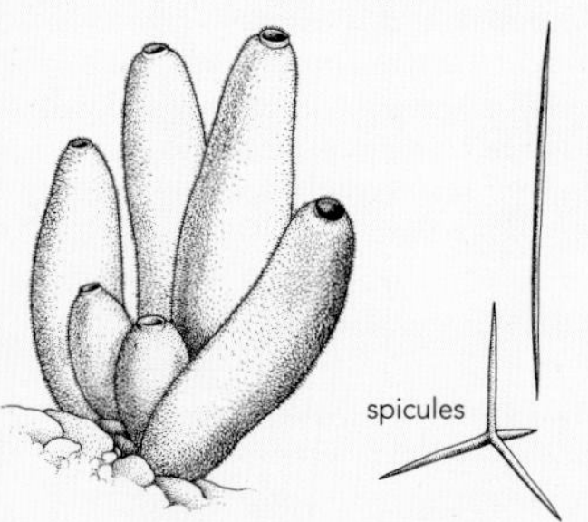

Calcareous sponge (1 in/2.5 cm)

PHYLUM CNIDARIA

The cnidarians display radial symmetry and their body plan is based on two layers of cells separated by a noncellular layer of fibers and proteins, the mesogloea. The diagnostic feature of these animals is the presence of the stinging cells, nematocysts, or cnidocytes, which vary from penetrating forms containing toxic solutions, to types which extrude threads that curl around the bristles and small appendages of the prey. The most toxic cnidarians are capable of killing humans, although the nematocysts, which are grouped on the tentacles, are primarily used to capture prey.

All the cnidarians have the same basic plan, with a mouth surrounded by tentacles that opens into a central body cavity. The mouth also functions as an anus and the undigested material, such as mollusk shells and crustacean exoskeletons, are voided via this central opening. The body plan of cnidarians is of two basic types, the polyp, a sedentary tubular animal with the mouth facing upward, and the pelagic or free-swimming medusae, in which the mouth generally faces downward. These body forms may alternate in the two generations of the same species or occur together in colonial forms.

All the 9000–10,000 species of cnidaria are aquatic, and the vast majority of species are marine, including the jellyfish, hydroids, sea anemones and corals. There are four classes of Cnidaria of which only one, the Hydrozoa, contains freshwater species, the remainder being marine. The Hydrozoa are the simplest of the cnidarians, numbering around 2000 species. The majority of hydrozoan polyps are generally of small size, although the genus *Branchioceri-anthus* found in depths of several thousand meters may reach 7 feet (2 meters) in length. The Scyphozoa number around 250 species and include some of the largest-known cnidarians, such as *Cyanea*, with a bell diameter of

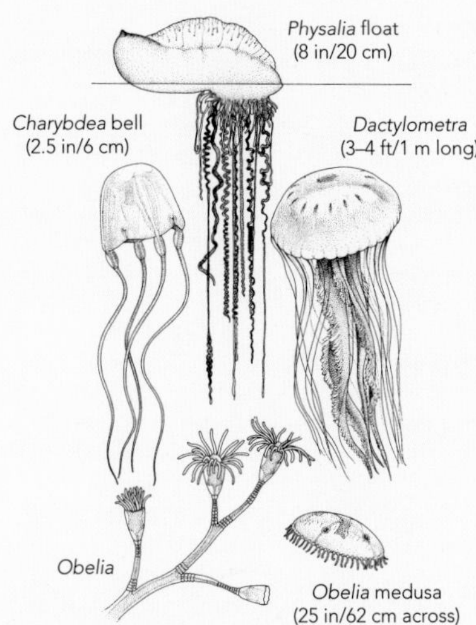

7–10 feet (2–3 meters) and up to 800 or more tentacles, each up to 60 meters (200 feet) in length. Most members of this group are free-swimming, medusoid forms. The cubomedusae, or box jellyfish, are basically cuboidal in form with the bell having four flattened sides.

The Anthozoa is the largest class with between 6000 and 8000 species of sea anemone and corals. They have a basic polyp body form in which the central gastric cavity is divided by septae that increase the surface area for digestion. The mouth is surrounded by eight branched tentacles in the Octocoralia, but six, or multiples of six, in the Zoantheria, the sea anemones (Actinaria) and true corals (Madreporia or Scleractinia). The most conspicuous and important members of this group are the hermatypic, reef-building corals, whose calcium carbonate skeletons form massive geological structures, such as the Australian Great Barrier Reef. The octocoralia include the sea pens, sea fans, whips and pipe corals in which the polyps are rather small, growing as a colony supported by a central, rodlike skeleton of gorgonin.

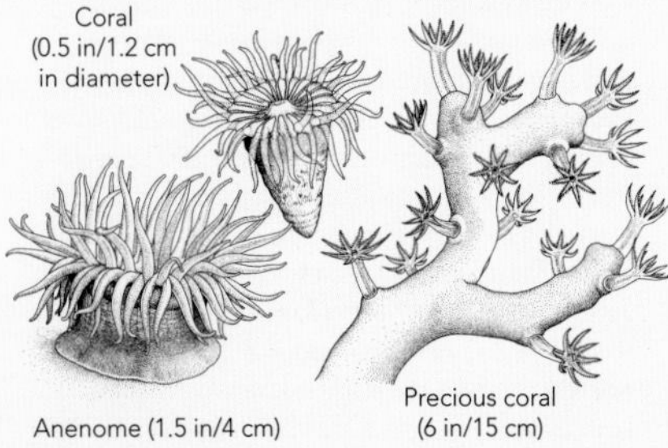

PHYLUM CTENOPHORA

The transparent and delicate sea gooseberries are a remarkable group of around 50 species of animals, many of which luminesce at night. They are biradially symmetrical and possess colloblasts, with similarities to the cnidarian nematocysts. They possess eight rows of fused cilia, forming a comblike structure which beats to provide the power for swimming. Most species are pelagic and *Pleurobranchia*, the typical sea gooseberry, has a spherical shape while, *Cestum*, the Venus' girdle, has an elongate and flattened body and can swim by lateral undulations of the body. *Cestum* includes some of the largest ctenophores, which may reach 3 feet (1 meter) in length.

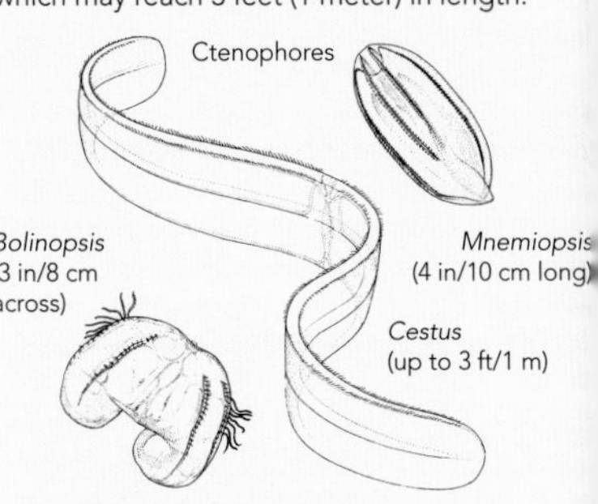

PHYLUM PLATYHELMINTHES

The 12,000–15,000 species of flatworms are not true worms, but flattened, bilaterally symmetrical, unsegmented animals which lack a true body cavity or coelom. The external epidermis is covered with cilia and the gut, if present, lacks an anus. Among the flatworms there are both parasitic and free-living forms. Twelve families of fluke are known to infect teleost or bony fish; five attack elasmobranchs; four, the ratfishes or holocephali; and only two are found in chondrostean fishes. There are four classes of platyhelminthes: the Turbellaria, or free-living flatworms; the Cestoda, or tapeworms; and the Monogenea and Digenea, or flukes.

PHYLUM NEMERTINEA

This small phylum of some 600 species of ribbon worms are, like the flatworms, creeping, burrowing animals with a ciliated epidermis.

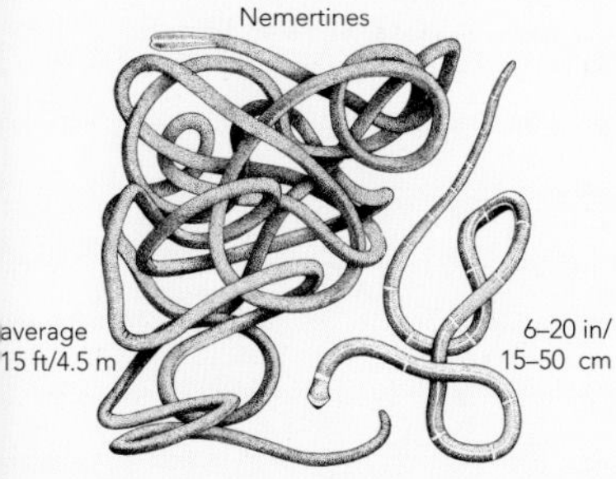

Most are littoral, some are found at abyssal depths and a few are pelagic. They are effective predators, catching and consuming their prey by means of a long proboscis, armed with sharply pointed stylets in the class Enopla, and lacking such structures in the Anopla. Most species are small, a few millimeters or centimeters in length, although the typical bootlace worm, *Lineus longissimus*, of temperate shores can achieve lengths of up to 16.5 feet (5 meters).

PHYLUM MESOZOA

A peculiar group of minute, wormlike animals, which includes around 50 known species, all of which are internal parasites of invertebrates. They include flatworms, nemerteans, annelids, mollusks and echinoderms.

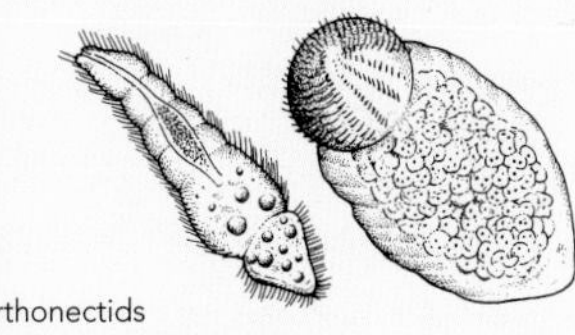

Orthonectids

PHYLUM GNATHOSTOMULIDA

A phylum of around 80 species of small – 0.02–0.04 inches (0.5–1 millimeter) – acoelomate worms which live between sand grains. They have an elongate body with a ciliated epidermis and feed on bacteria and fungi, which are seized by a pair of toothed, jawlike structures. They are hermaphroditic with fertilization occurring when the male penetrates the female body wall with a copulatory organ. The eggs are shed by rupture of the female's body wall.

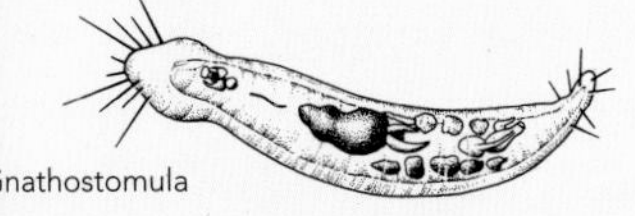

Gnathostomula

SUPER-PHYLUM ASCHELMINTHES

The super-phylum Aschelminthes are wormlike animals which lack definite heads but display some degree of radial symmetry at their anterior end. They have a complete digestive system with mouth and anus, a simple reproductive system, but no respiratory and circulatory systems. They have a reduced number of cells making up the body, and many groups have a well-developed scleroprotein cuticle on the surface of the epidermis.

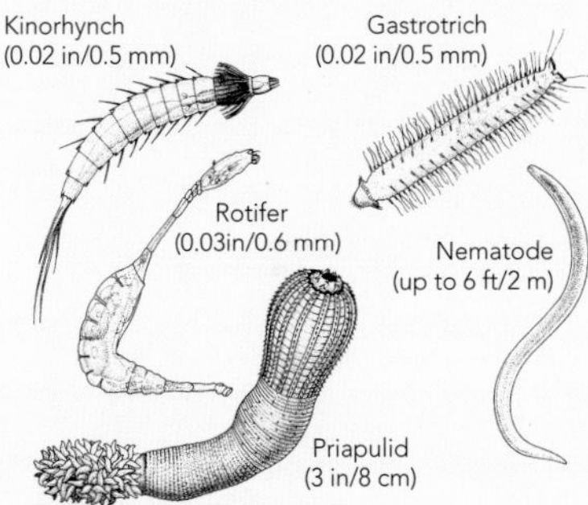

PHYLUM NEMATODA

Although only some 10,000 species have been described, it has been estimated that as many as 500,000 species may exist. Nematodes are small and inconspicuous, and there are immense numbers of individuals. The majority are free-living, distributed from the tropics to the poles and from the tops of mountains to the abyssal depths of the ocean. The phylum is divided into two classes: the Aphasmida, which lack chemoreceptive organs, and the Phasmida, which have a pair of chemoreceptive organs on either side of the tail. The former include the majority of marine and fresh-water species, while the latter includes the majority of parasitic forms.

PHYLUM ROTIFERA

These tiny animals number around 1500 species and are comparable in size to the ciliated protistans with which they often live and compete. They have a corona, or crown of cilia, which, when it is in motion, resembles a spinning wheel. Only a few species are marine, but like the nematodes, they are cosmopolitan in distribution. These animals are sedentary and reproduce both sexually and parthenogenically, females producing eggs that hatch to produce more females.

PHYLUM GASTROTRICHA

This phylum contains only about 150 species of mainly freshwater animals, with a small number of species occurring interstitially, in intertidal sands. They are generally scaly or spiny on the dorsal surface, with cilia for locomotion beneath. Most species feed on organic debris, algae, protista and bacteria which are swept into the terminal mouth by four groups of cilia around the anterior end of the animal. They generally have a short lifespan of between three and 21 days, and are hermaphroditic. Four or five eggs of two types are produced: thick-shelled ones, used to avoid unfavorable conditions, and thin-shelled ones, which hatch in a few days, the animals maturing in two days.

PHYLUM KINORHYNCHA

Like the rotifers and gastrotrichs, the Kinorhynchs are microscopic, less than a millimeter in total length. The body is generally elongate, consisting of a head, neck and 11 body segments. The majority of the 100 known species are found in ocean muds in shallow waters. The dorsal surface bears spines, while the head has 5–6 circles of spines and an oral cone of stylets. Detrital material is sucked in by contractions of the muscular pharynx, and the head can be retracted into the body. Like the gastrotrichs, the Kinorhynchs have a pair of posterior adhesive organs allowing them to attach temporarily to the surface of the substrate.

PHYLUM NEMATOMORPHA

The hair worms are a small phylum of some 230 species of extremely slender worms, which are brown or black in color. Most species are terrestrial, but a single genus, *Nectonema*, occurs in the marine environment where it parasitizes crabs in its juvenile stages. They may reach 3 feet (1 meter) in length but are usually less than 0.04 inches (1 millimeter) in diameter. The sexes are separate; males entwine around the females to fertilize the eggs, which hatch as larvae that then enter the host.

PHYLUM ACANTHOCEPHALA

The spiny-headed worms are a group of 800 or so species that are all internal parasites of vertebrate guts, and they have an intermediate, arthropod host. Fish are the most common vertebrate host, and the majority of these worms are around 0.8 inches (2 centimeters) in length. They may be present in large numbers, and over 1000 have been reported from a single seal. The sexes are separate and fertilization occurs within the female. The eggs develop into an acanthor larva that, when ingested by a crustacean intermediate host, hatches and the larva bores through the gut wall using its hook-bearing rostellum. The animal completes its development when the intermediate host is eaten by the vertebrate primary host.

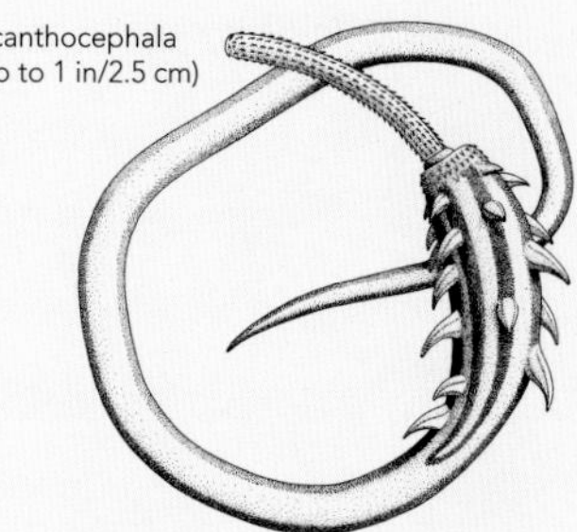
Acanthocephala (up to 1 in/2.5 cm)

THE COELOMATE ANIMALS AND METAMERIC SEGMENTATION

All the remaining members of the animal kingdom possess, at least during some stage of their life history, a true body cavity or coelom. There are five major phyla (Annelida, Arthropoda, Mollusca, Echinodermata and Chordata), with several minor phyla. An important structural adaptation is the development of segmented body form, with replication of all structures in each segment along the length of the body.

PHYLUM ANNELIDA

This phylum of some 9000 species of worms is recognizable by the ringlike divisions of the body, which represent the individual segments. These animals resemble an elongate tube with terminal mouth and anus, and a straight digestive tract between. The body cavity in each

Mud-dwelling polychaetes

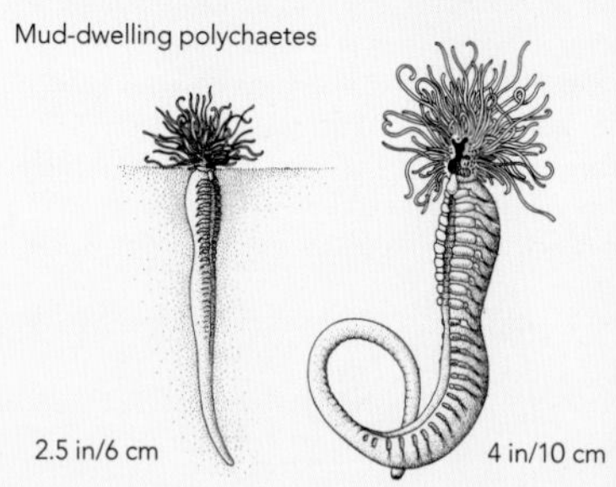
2.5 in/6 cm

4 in/10 cm

segment is separated from its neighbors by a membranous septum. The fluid-filled cavities of the body form a hydraulic skeleton; contraction of the longitudinal muscle fibers causes shortening and thickening of the segment and hence of the animal as a whole, while contraction of the antagonistic circular muscles causes elongation and narrowing of the segment. By alternating contractions of these muscles in different segments, the animal can pull itself forward in a burrow or, by means of its appendages and bristles, move over the surface of the substrate. There are four classes of annelids: the polychaetes, which include some 5500 species of marine worms; the predominantly freshwater and terrestrial oligochaetes; the Hirudinea, with 500 species of marine, freshwater and terrestrial leeches; and the Myzostomaria, a group of small, flattened, disc-shaped worms that are parasites of crinoids.

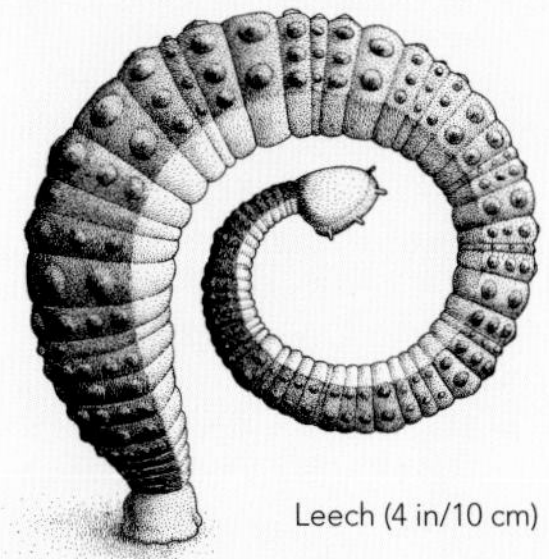

Leech (4 in/10 cm)

PHYLUM ARCHIANNELIDA

The archiannelids, once thought to be primitive members of the Annelid phylum, are now believed to represent a group of unrelated families. They are generally interstitial, living in surface mud and in the splash zone. Lacking bristles, parapodia and any signs of external segmentation, they have external cilia and simply arranged heads.

PHYLUM ARTHROPODA

The phylum Arthropoda is the largest group in the animal kingdom, with an estimated million species and over 800,000 described species at the present time. In the arthropods, specialization of regions of the body for particular purposes is a major characteristic of the group. The main feature distinguishing arthropods, however, is the tough semirigid exoskeleton consisting of chitin, a polysaccharide and protein mixture,

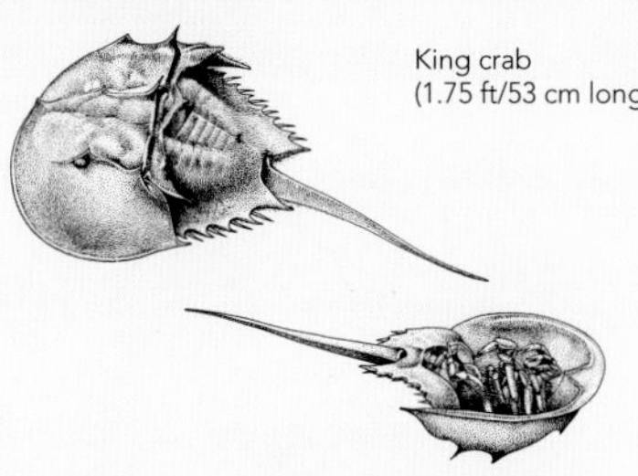

King crab (1.75 ft/53 cm long)

with a thin, impermeable external cuticle. In marine arthropods, such as crabs, the cuticle is thick and calcified. The jointed appendages can be moved independently of one another and of the body, by internally inserted muscle blocks. The group is divided into four subphyla, one of which, the Trilobomorpha, is an entirely fossil group of primitive animals, the trilobites. The remaining three subphyla include the Chelicerata, the spider's scorpions and ticks on land, and the king crabs; and Pycnogonida, or sea spiders of the marine environment. The subphyla Crustacea and the Uniramia are distinguished on the basis of their limb anatomy. The Uniramia are represented in the marine environment only by a few insects, such as sea skaters.

In the subphylum Crustacea, all but a few species are aquatic, with the vast majority being found in the ocean. About 31,300 species of crustacea are presently known, and these are generally divided into seven classes of which the Malocostraca is the most advanced and numerous, containing some 20,000 species. The classes Branchiopoda; Ostracoda and Copepoda are generally small planktonic grazing forms, while the class Cirripedia, or barnacles, are a worldwide group and are the only nonparasitic crustaceans to have adopted a sessile mode of life. The class Mystacocarida is a minor class, related to the copepods and barnacles, and including a single genus, *Derocheilus*, first discovered in 1943. The class Branchiura is a group of around

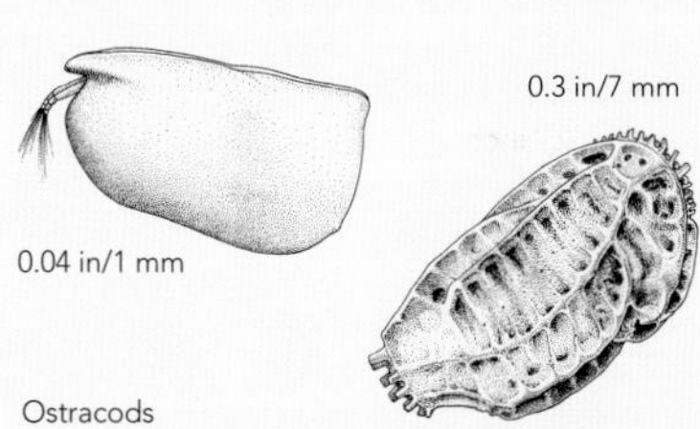

Ostracods

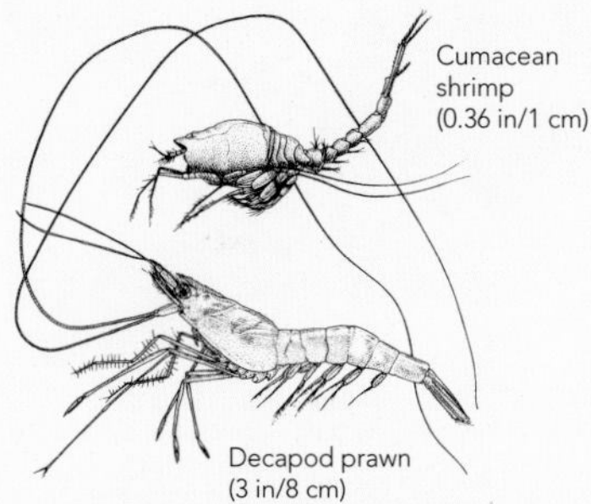

75 species of ectoparasitic blood suckers, living on fish and some amphibians. The class Malacostraca are entirely aquatic, and the most successful group is the superorder Eucarida, characterized by a well-developed carapace and including such well-known marine organisms as the krill, which form the food of whales in the Antarctic. The decapods are specialized either as swimmers, the shrimps and prawns; or crawlers, the bottom-dwelling crabs, crayfish and lobsters.

PHYLUM PRIAPULOIDEA

This phylum contains nine species of small burrowing animals which live in mud or sand in cool- or cold-water areas down to depths in excess of 23,000 feet (7000 meters). They range in size from the microscopic *Tubiluchus*, which is only 0.2 inches (0.5 millimeters) in length, to some species of *Priapulus*, which can reach 8 inches (20 centimeters). The body is somewhat barrel-shaped with an eversible proboscis armed with spines used to catch slowly-moving worms and other soft-bodied animals.

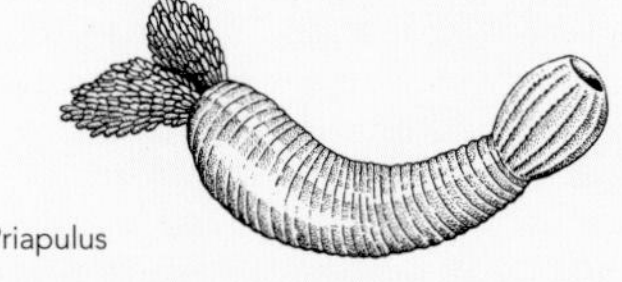
Priapulus

PHYLUM SIPUNCULOIDEA

This is a widespread group of around 350 species of active burrowers, which range in size from 0.08–29 inches (2–720 millimeters) and are cylindrical with no external segmentation. They have a short protrusible proboscis which ends in an oral disk of frilly tentacles. The larval forms of sipunculids are planktonic (aiding dispersal) and include a typical trochophore larva resembling that of annelid worms.

PHYLUM ECHIUROIDEA

The 100 or so species show some affinity to the sipunculid worms and, like them, are sausage-shaped animals that live buried in mud or sand. They are readily recognizable by the extremely long proboscis, which is non-retractable. In several genera, such as *Bonellia* and *Echiurus*, the minute males are parasitic, living permanently attached to the surface of the female. Like the sipunculid worms, these animals have a trochophore larva and are considered to be related to the annelids.

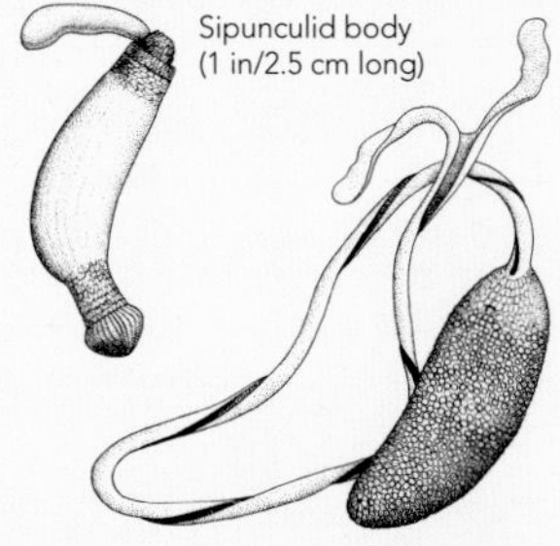

Echiuroid (2.25 in/6 cm across)

PHYLUM POGONOPHORA

The pogonophora inhabit marine environments below depths of around 330 feet (100 meters) and are threadlike animals, up to 1.2 feet (35 centimeters) in length, but usually no more than 0.04 inches (1 millimeter) in diameter. Only discovered in 1900, the 100 or so species are now known to be widely distributed. They have no mouth or digestive tract and are believed to digest their food externally, absorbing the products through the tentacles. The elongate body ends in an opisthoma, which bears bristlelike chaetae and serves to anchor the animals in their burrows.

PHYLUM TARDIGRADA

The water bears are a peculiar group of cosmopolitan animals generally less than 0.02 inches (0.5 millimeters) in length. Their cylindrical bodies are equipped with four pairs of short, stubby legs, which end in bunches of four or eight hooks. The external surface is covered with a cuticle, which is molted periodically, and mating only occurs during the time of the molt. Some species are marine, living among sand particles, but the majority are semiaquatic, living in the water film associated with terrestrial vegetation.

PHYLUM PENTASTOMIDA

The pentastomids are a group of around 90 species of blood-sucking parasites which are believed to be related to the brachyuran crustacean fish parasites. As adults, they inhabit the respiratory tract of reptiles, mammals and birds. Common, intermediate hosts of pentastomids include fish and small mammals.

PHYLUM MOLLUSCA

This is the second largest phylum of the animal kingdom, with around 100,000 living species and some 35,000 described fossils. The group is diverse and the vast majority of species are marine. The group evolved in the ocean realm, having only a few gastropods (snails) and bivalves in freshwater and even fewer gastropods on land. Mollusks are a group of bilaterally symmetrical animals with a reduced coelom in which the well-developed head, with associated sense organs, is continuous with a muscular structure, the foot. The body organs form a visceral mass on the dorsal surface, and the foot is modified in different groups for creeping, burrowing and digging.

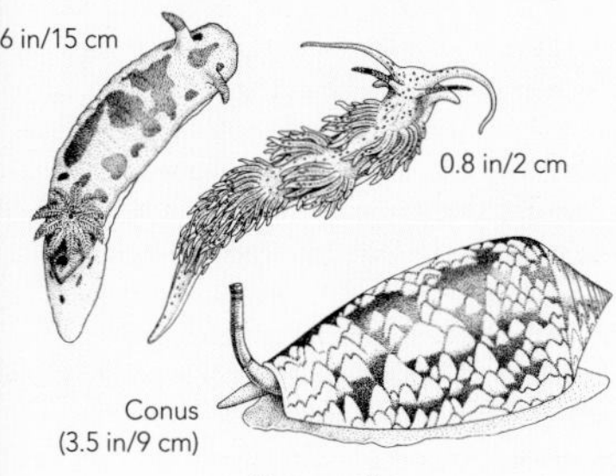

Gastropods

The mollusks have a well-developed and often beautifully colored shell which is secreted by the mantle and consists of a protein matrix reinforced by crystalline calcium carbonate in the form of calcite or aragonite. The enormous diversity in body form is reflected in the classes of mollusks, each of which is based on a somewhat different body plan.

The Bivalvia (oysters and clams) have a shell of two halves hinged together while the Gastropoda (snails), with some 75,000 living species, have a single shell or have lost it altogether. The Cephalopoda, squid, octopuses and cuttlefish, are soft-bodied with a generally active mode of life and well-developed sense organs. The class Monoplacophora, though

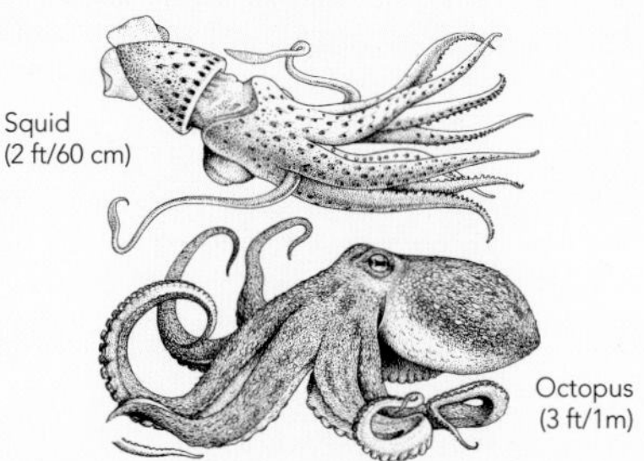

Cephalopods

extinct, is known from a few species in the genus *Neopilina* that occur in abyssal depths of between 6600 and 23,000 feet (2000 and 7000 meters). The class Polyplacophora are the chitons or mailshells, easily recognizable by their shell formed of eight separate plates, while the Scaphopoda are a small group of tubular mollusks which generally live buried in sand or mud with the tip of the shell protruding above the surface. The Aplacophora are a group of peculiar, shell-less, worm-shaped mollusks, which live among corals and hydroids on which they feed.

BRYOZOA, BRACHIOPODA AND PHORONIDA

The Lophophorate phyla, the bryozoa, or sea mosses, brachiopods and phoronids, are three minor phyla of coelomate animals which all possess a lophophore – a cluster of tentacles used for gathering food. All are sessile, with the body divided into three sections, the middle section of which is modified to form the crown of tentacles, and all possess a U-shaped digestive tract with anus opening on the upper surface. Bryozoa number some 4000 species, and this is the most abundant and widespread lophophorate phylum occurring on virtually any solid substrate along the shore, fouling marine structures and ship hulls.

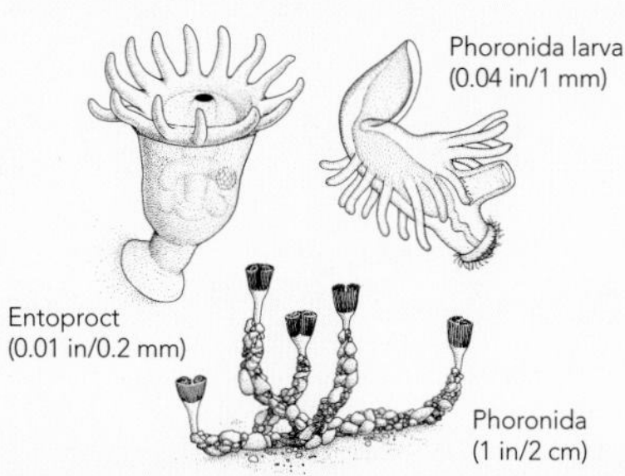

The Brachiopoda, or lampshells, are an entirely marine group of lophophorates that superficially resemble bivalve mollusks since they have a shell comprised of two horny valves impregnated with calcium carbonate. The phylum Phoronida, or horseshoe worms, comprise one of the smallest phyla of around 10 marine species which live in chitinous tubes buried in sand in shallow water.

PHYLUM ECHINODERMATA

This phylum contains approximately 6000 living species of exclusively marine animals. The group has an extensive evolutionary history extending back to the Cambrian and including over 20,000 described fossil species. As a group, they are immediately recognizable by their pentaradial (five-rayed) symmetry, although some of the starfishes, for example, have numerous arms in multiples of five, while the five-rayed symmetry of some burrowing sea urchins is difficult to see at first glance. In sea urchins and starfish, the mouth generally faces downward; in others, such as the feather stars and sea lillies, the mouth faces upward. One distinctive feature of the echinoderms is their complex water vascular system, a network of fluid-filled tubes which connect the tube feet that protrude outside the calcium carbonate skeleton and which are used for locomotion, burrowing, and in some groups, feeding. The skeleton consists of a 'test', or box of plates, joined edge to edge and perforated by a series of pores through which the tube feet and respiratory papullae are extended. The surface of the test is covered by the animal's epidermis and specialized structures; the pedicellariae may be used to remove settling larvae from the surface of slowly-moving sea urchins and some starfish.

There are five classes of echinoderms: the Crinoidea, which includes the sea lillies, most of which occur in deep waters, and the feather stars, which are abundant in shallow tropical waters; the Asteroidea, or familiar sea stars, which are widely distributed in shallow water environments; the Ophiuroidea, or brittle stars, another widely distributed group; the Echinoidea, or sea urchins; and the Holothuroidea, or sea cucumbers.

Asteroids

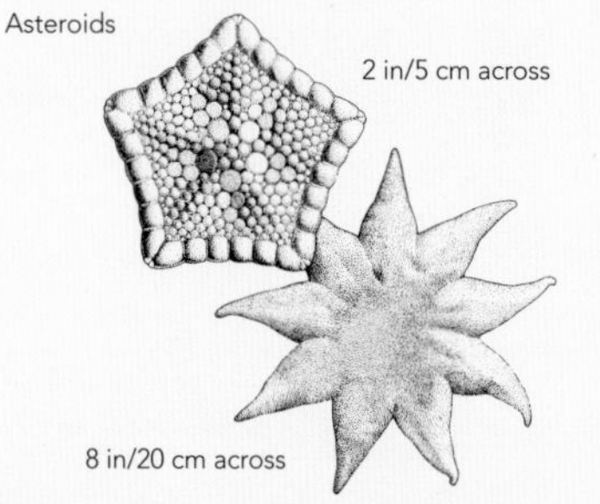

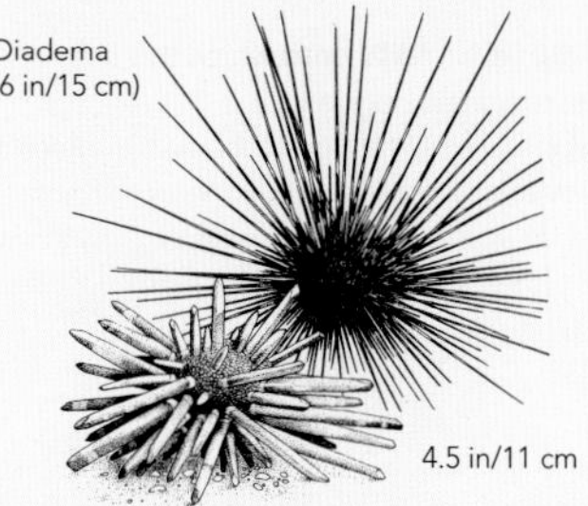

Echinoids

PHYLUM CHAETOGNATHA

This is a group of 50 species of marine organisms which are extremely common in the oceanic plankton, all of which display a rather uniform body plan. They are recognizable by their elongate, arrow-shaped body, paired lateral fins and a single tail fin. The anterior mouth has strong grasping spines, and these animals are among the most important carnivores of the plankton community, feeding particularly on copepods. They are generally small, around 0.8 inches (2 centimeters) in length, and the head is comparatively small, most of the animal consisting of a trunk section with paired lateral fins which terminate just in front of the single tail fin. Fertilization is internal, and some species carry their eggs, others release them into the plankton, while the benthic genus *Spadella* attaches them to rocks and seaweeds in the littoral zone.

PHYLUM HEMICHORDATA

This phylum of around 100 species displays two distinct body forms: the class Enteropneusta, or acorn worms, are elongate wormlike animals, while the Pterobranchs are sessile, tube-dwelling, deep-water animals of peculiar body form. They are recognizably distinct from the preceding animal phyla in having the body clearly divided into three regions: an anterior protosome, modified into a short proboscis in the acorn worms and as a shield-shaped structure which secretes the tube in pterobranchs;

a mesosome, or collar, into which the acorn worm can retract the proboscis and which in pterobranchs is strikingly modified into a series of hollow arms bearing ciliated tentacles used in feeding; and a metasome or trunk, which is elongate in the acorn worms and short and barrel-shaped in pterobranchs.

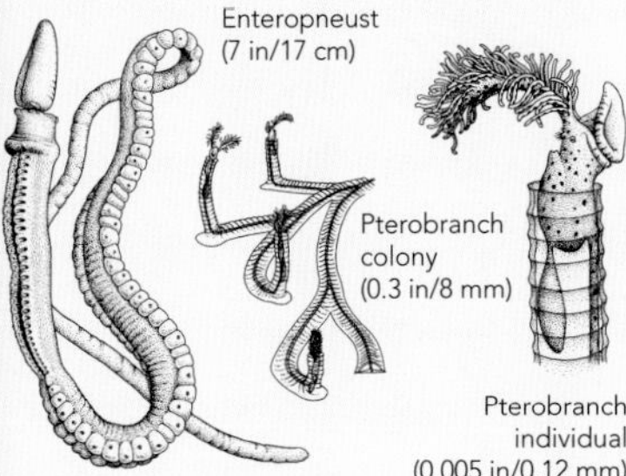

Hemichordates

PHYLUM CHORDATA

The phylum chordata, which includes the vertebrates, is the largest deuterostome group of animals and the youngest in evolutionary terms. While the fossil record for invertebrate phyla extends back some 1600 million years, fossils of the earliest vertebrates are only 500 million years old. Chordates are, however, one of the major animal phyla dominating land, air and water, although they do not exceed the arthropods, either in terms of numbers of species or individuals, they far outstrip them in terms of total biomass and ecological importance. Three chordate characters are considered diagnostic of this phylum. The first is the presence, at least during some stages of their life history, of an axial skeleton either in the form of a stiffened rod, the notochord, or in higher forms a jointed bony structure, the vertebral column. Secondly, the presence of clefts or gill slits, and thirdly, the possession of a dorsal, hollow nerve tube, expanded at the anterior end to form a brain, often protected by a boxlike structure, the cranium. In addition, most chordates possess a post-anal tail which is flexible and muscular and, in most aquatic forms, is used for locomotion. The circulatory system consists of a closed tubular network with a muscular pump, the heart.

SUBPHYLUM UROCHORDATA

Urochordates, or tunicates, are common marine animals numbering some 1300 species,

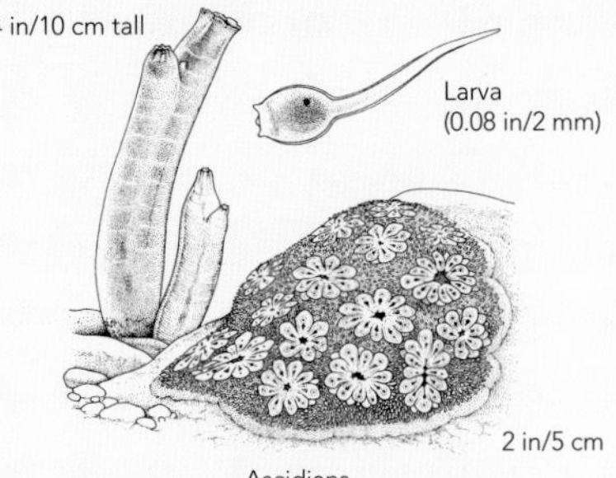

Ascidians

which as adults bear little resemblance to other chordates and are placed by some authors in a separate phylum, Tunicata. The most common and widespread group, the class Ascidiacea, or sea squirts, are adapted to a sessile mode of life, resembling a barrel bearing two siphons through which water is drawn for both respiration and feeding. The water is passed through a net, and food particles are trapped in mucus and rolled toward the opening of the digestive tract. Water, which has been filtered, passes through the net and out of the body of the animal via the atrial siphon. Only during development do these animals display the chordate features of a notochord or axial skeleton, a dorsal hollow nerve cord and gill clefts. Two other classes of ascidean, the Thalacea and the Larvacea, are both modified for a pelagic existence. The thalaceans, a small group of only six genera, resemble small barrels with one siphon at each end, so that water passes through the length of the animal.

In contrast, the larvaceans retain the tadpolelike body form throughout life; the tail is long and extends at right angles from the dorsal surface.

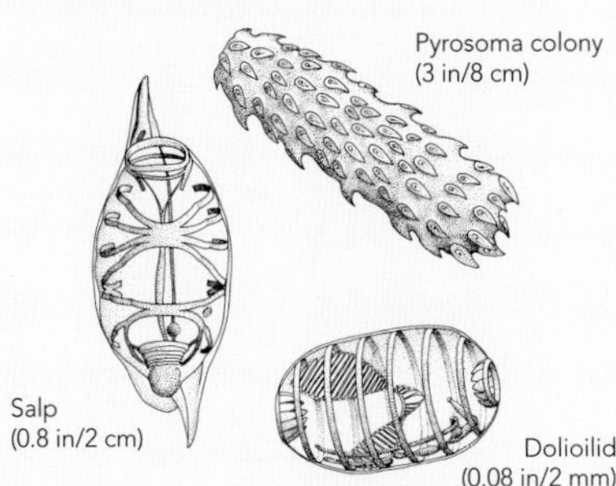

Thaliaceans

SUBPHYLUM CEPHALOCHORDATA

Only two genera comprise this subphylum, *Branchiostoma* and *Asymmetron,* distinguished by the fact that in *Asymmetron* the gonads are restricted to the right side of the body, and in *Branchiostoma* they are paired. Adults are bottom-dwelling, filter-feeding animals, using a net of similar construction to that of tunicates for trapping fine, suspended particles. These animals are generally small, around 1.6–3 inches (4–8 centimeters) in length, and the characteristic notochord and dorsal hollow nerve cord of the chordates are present throughout life, as is the post-anal tail. The muscles, nerves and blood vessels of the body wall are segmentally arranged, and the alternate contractions of the myotomes, or muscle blocks on each side of the body, cause lateral movements, allowing the animal to swim in a fishlike manner.

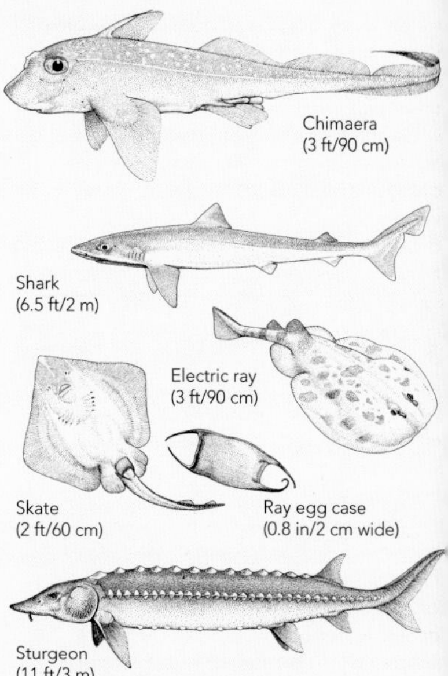
Chimaera (3 ft/90 cm)
Shark (6.5 ft/2 m)
Electric ray (3 ft/90 cm)
Skate (2 ft/60 cm)
Ray egg case (0.8 in/2 cm wide)
Sturgeon (11 ft/3 m)

SUBPHYLUM VERTEBRATA

The most primitive vertebrates were jawless, fishlike animals, which appear in the fossil record during the Ordovician and are related to the extant hagfish, and lampreys. Modern classifications now recognize three major groupings of extant fishes and fishlike animals: the Agnatha, hagfish, and lampreys; the Elasmobranchiomorphi (Chondrichthyes), sharks, rays and ratfishes, with a cartilaginous skeleton; and the Osteichthyes, or bony fishes. In addition, an extinct fossil group, the Placodermi, includes early, heavily armored, bottom-dwelling forms considered either as a subclass of the Elasmobranchiomorphi or as a separate class.

Class Agnatha

The lampreys and hagfish lack a vertebral column and paired fins but have a notochord like the cephalochordates. They also lack the jaws of the higher vertebrates and the mouth is circular, terminal and armed with hooklike structures. Lampreys have a muscular rasping tongue which is used to scrape the surface of the fish to which they attach, feeding on blood and tissue. In contrast, the hagfish seem to be scavenging animals feeding on dead and dying animals. Sea lampreys may attach to large basking sharks and whales, but unlike the hagfish, which are entirely marine, lampreys return to freshwater to breed, where the larvae, which resemble cephalochordates, undergo a lengthy period of development.

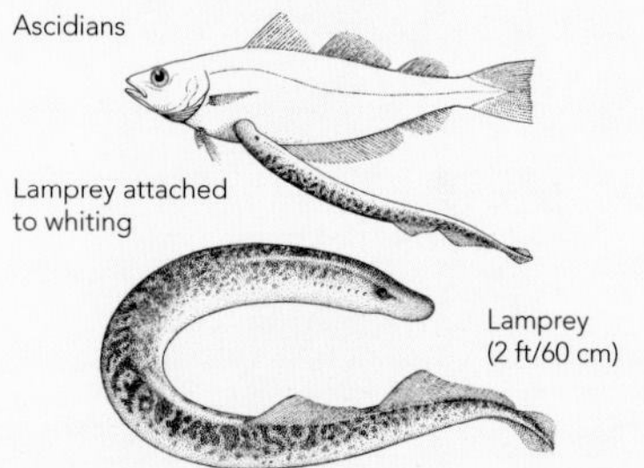
Ascidians
Lamprey attached to whiting
Lamprey (2 ft/60 cm)

Class Elasmobranchiomorphii (Chondrichthyes)

This group of animals is distinguished from the higher fishes by the presence of a cartilaginous, rather than a bony, skeleton. The class is divided into two major groupings, the Holocephali, or ratfishes; and the Elasmobranchii, which includes more than 2000 species of sharks, skates and rays. The ratfishes are a small group of peculiar fishes in which the gill slits open externally through a single opening. The tail is long and whiplike, and these animals swim by means of enlarged pectoral fins. The head is large, and the jaws have strong, platelike teeth for crushing the mollusks on which they feed.

Modern elasmobranchs have well-developed jaws generally armed with sharp teeth and capable of powerful jaw closure although often without the capacity to chew. The great white shark, *Carcharodon carcharias*, has serrated teeth and is a voracious predator of fish, including other sharks and large vertebrates, such as seals, sea lions and dolphins. Most sharks are solitary, although the thresher shark, *Alopias*, hunts in packs using its long whiplike tail to drive the herring, mackerel or pilchards into a compact shoal. The basking shark, *Cetorhinus maximus*, which grows to a length of 35 feet (10.5 meters), and the whale shark, *Rhincodon*, which grows to a length of 60 feet (18 meters), feed by swimming around with their mouths open, filtering small plankton from the surrounding water.

The skates and rays are specialized for a bottom-dwelling existence, being flattened forms with enlarged, lobelike pectoral fins which form the typical wings. The tail fin is not generally used in locomotion and, in many species, it is reduced to a long, whiplike structure armed with spines. Most species have a teeth for grinding, and *Myliobatis*, the eagle ray, feeds almost exclusively on clams and oysters, which are crushed between a grinding mill of flattened barlike teeth.

Herring
(12 in/30 cm)

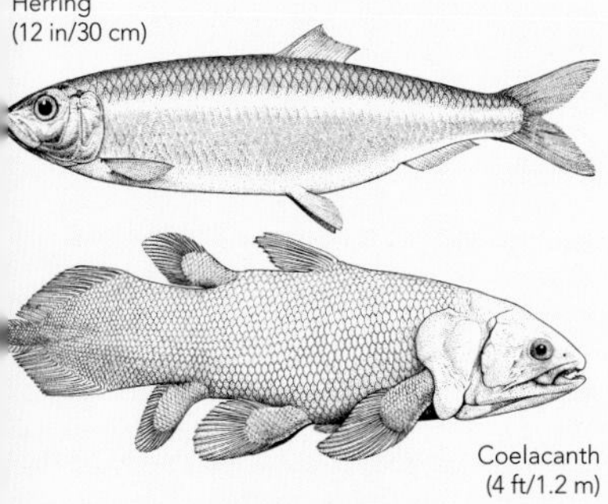

Coelacanth
(4 ft/1.2 m)

Class Teleostomii (Osteichythes)

Two major groupings of bony fish are recognized: the Sarcopterygii, or lobe-finned fishes, with only seven living species, and the Actinopterygii, or ray-finned fishes, which number in excess of 30,000 species. The lobe-finned fish are important as the ancestral group from which the land vertebrates evolved, and the living *Latimeria*, or coelacanth, of the Indian Ocean, and the freshwater lungfish, of tropical regions, are the only living representatives. *Latimeria* is a large fish reaching more than 200 lbs (90 kg) in weight and feeding on other fish, which it usually swallows whole.

Most of the 30,000 living marine and fresh-water fishes are members of the Teleostei, of which the herringlike fishes are the most primitive, having an elongate fusiform body with pectoral and pelvic fins placed toward each end of the animal, a median dorsal and anal fin which act as stabilizers during swimming and a large caudal or tail-fin which provides the main propulsive force. Alternating contraction of the muscle blocks on each side of the animal sweep the tail from side to side and, in more advanced teleosts, this lateral movement involves only the tail fin itself, the remainder of the body being held relatively rigid. The diversity of body form in this group of fishes is staggeringly large, ranging from long, eellike burrowing forms, globular slow-swimming forms, and flattened, bottom-dwelling forms, to high-speed swimmers such as tuna with a torpedo-shaped body and greatly enlarged pectoral fins.

Class Reptilia

Although this group of animals dominated the marine environment during the Mesozoic era, they are now restricted to seven species of marine turtle, several hundred species of sea snakes, one species of marine iguana and a few estuarine crocodilian species. The turtles have horny beaks and modified front limbs as paddles, which are the main source of power for locomotion, the rear flippers being used only for steering. Sea turtles have reduced the weight of the skeleton and assumed a flattened, streamlined shape. They come ashore to breed, laying eggs in excavated nests at the head of beaches.

Leatherback turtle
(10 ft/3 m)

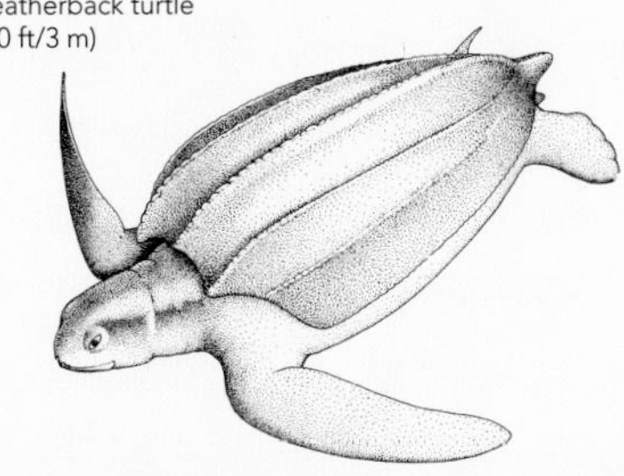

The marine iguana, *Amblyrhynchus*, of the Galapagos Islands, is really only an amphibious lizard spending a considerable portion of its time on land. Like seabirds, *Amblyrhychus* has salt-excreting glands in the nostrils, in contrast to turtles, which excrete excess salt through ducts in the eye socket.

Sea snakes are a diverse group of highly venomous marine animals, related to the terrestrial cobras and kraits but the tail is laterally compressed, forming a paddle. *Pelamis* is completely pelagic, feeding on small fish, while many of the smaller-headed species feed on fish eggs. Most sea snakes are ovoviviparous, the eggs hatching inside the female's reproductive tract, and the three to eight young emerging as fully formed juvenile snakes.

Class Aves

Seabirds range from wading species found around the fringes of the world's oceans to truly oceanic species, such as petrels and albatrosses, which come on land only to nest. All seabirds feed on marine organisms, with the waders probing mud and silt to extract small worms and crustaceans, and diving birds feeding on fish in surface waters. The penguins have lost the ability to fly and have modified the wings to form flippers for swimming underwater in pursuit of their fish prey. Other diving birds, such as the true divers and cormorants, use their feet for underwater swimming. Like the turtles, seabirds must nest on land, and seabird colonies are generally concentrated in suitable coastal areas. Kittiwakes and gannets nest on rocky cliffs, while many gulls and terns nest on the ground in sand-dune areas, boobies and albatross in trees on isolated oceanic islands, and penguins on ice-free areas of the Antarctic and other southern continents.

King penguins (3 ft/1 m)

Class Mammalia

Marine mammals belong to a number of different groups, with the polar bear and sea otters being representatives of the largely land-based order, the Carnivora. They display fewer adaptations to a marine existence than do the more specialized seals, the Pinnepedia, to which they are related. The three groups of true marine mammals are: the Pinnipedia, or seals; the Sirenia, or sea cows; and the Cetacea, the whales and dolphins.

There are only four living species of dugong and manatees. A fifth, Steller's sea cow, was hunted to extinction in the late 18th century. Distantly related to elephants, these animals have lost the hind limbs completely and modified the forelimbs to form strong paddles. The tail is flattened like the flukes of whales, and the animals are herbivorous, being confined to shallow water areas, estuaries and large river systems of the tropics and sub-tropics where they graze on seagrasses.

The seals, sealions and walruses are a diverse group of mainly fish-eating species that possess streamlined bodies with reduced or absent hair cover and insulation provided by a layer of fat beneath the skin. The walrus, with its elongate downwardly pointing tusks, is an Arctic animal adapted to feeding on mollusks.

The most highly adapted marine mammals are the whales and dolphins with streamlined body, no visible neck, well-developed front flippers, and flattened and expanded tail flipper or flukes. Like the seals, they are hair-less and posses a thick layer of fat or blubber below the skin for insulation. Whales give birth in the water, and the baby whales are born tail first and nudged to the surface to take their first breath. The toothed whales are active predators, pursuing fish, squid and, in the case of killer whales, seals and penguins. The baleen whales lack teeth but have plates of baleen set in the upper jaw, which act as sifts, straining out the krill as the animals swim through the plankton with their mouths agape. These animals reach a large size, the smallest being 17 feet (5 meters) when fully grown. The largest, the blue whale, can reach more than 108 feet (33 meters) in length, the largest mammal to have ever lived on Earth.

Page numbers in *italic* refer to picture captions; **bold** numbers refer to major references.

PICTURE CREDITS

6 Reto Stöckli, Robert Simmon/*NASA GSFC*
10 B. Chouet/*USGS*
16 Ric Ergenbright/*Corbis*
19 Royalty free/*Corbis*
20 Science Pictures Limited/ *Science Photo Library*
21 Earth Satellite Corporation/ *Science Photo Library*
22 Yann Arthus-Bertrand/*Corbis*
24 JSC/*NASA*
25 JSC/*NASA*
29 Alfred Pasieka/*Science Photo Library*
31 Royalty free/*Corbis*
32 Jeff Rotman/*Nature Picture Library*
34 Jacques Descloitres, MODIS/*NASA GSFC*
36 Galen Rowell/*Corbis*
38 British Antarctic Survey/ *Science Photo Library*
41 Robert Estall/*Corbis*
44 JPL/*NASA*
46 Hal Pierce/GSFC/*NASA*
50 MODIS Ocean Group/*NASA GSFC, University of Miami*
52 SeaWiFS Project/*NASA GSFC, ORBIMAGE*
54 RSD/*NASA GSFC*
57 Ray Soto/*Corbis*
58bl&br TOPEX/Poseidon/ *NASA JPL*
61 Teobaldo Dioses/*NOAA*
62 Bettmann/*Corbis*
64 Archivo Iconografico, S.A./ *Corbis*
65 Bettmann/*Corbis*
66 *Mary Evans Picture Library*
67 The Stapleton Collection/ *Bridgeman Art Library*
68 *NOAA Central Library*
69tr *NOAA*
69c Steve Nicklas, NOS, NGS/ *NOAA*
70b OAR/National Undersea Research Program (NURP), Univ. of Hawaii/*NOAA*
73tl *NOAA*
73tr Randy Wilder/Monterey Bay Aquarium/*NASA JPL*
76 Family of Captain William M. Scaife, C&GS/*NOAA*
77 Peter Stirling/TAXI/ *Getty Images*
78 Royalty free/*Getty Images*
79 George D. Lepp/*Corbis*
80 The Mariners' Museum/ *Corbis*
83 Peter Scoones/*Science Photo Library*
85 Stephen Frink/*Corbis*
86 Scott Kerrigan/*Corbis*
87 Guido Alberto Rossi/The Image Bank/*Getty Images*
88 Nik Wheeler/*Corbis*
89 Sergio Dionisio/ Oceanfashion Pictures/ *swedishmatchtour.com*
90 Gary John Norman/Stone/ *Getty Images*
91 Oli Tennent/Stone/ *Getty Images*
92 Kaora Soehata/*PPL*
94 David Lees/*Corbis*
96 National Railway Museum, York/HIP/*TopFoto.co.uk*
97 Michael Setboun/*Corbis*
98 Kevin Schafer/Stone/ *Getty Images*
99 Christian Humphries
100–1 SeaWiFS/*NASA GSFC*
104bl Dr Anthony R. Picciolo/ *NOAA NODC*
104bc Constantinos Petrinos/ *Nature Picture Library*
106br Douglas P. Wilson, Frank Lane Picture Agency/*Corbis*
105t Stephen Frink/*Corbis*
105b Constantinos Petrinos/ *Nature Picture Library*
106 Nat Fain/*Natural Science Photos*
109 Mohammed Al Momany/ *NOAA*
110 Mohammed Al Momany/ *NOAA*
112 Thanks to Jerry D. Vandergriff, Anchorage
113 Jamie Hall/*NOAA*
115t Mark Carwardine/ *Still Pictures*
115b Stephen Frink/*Corbis*
116 Jeff Rotman/*Nature Picture Library*
117 Brandon Cole/*Nature Picture Library*
118 Joe Martin/*USWS*
119 Mark Newman/*Science Photo Library*
120 Gunter Ziesler/*Still Pictures*
121 Constantinos Petrinos/ *Nature Picture Library*
123 Stan Butler/*NOAA*
126 Captain Budd Christman/ *NOAA*
129 Pete Oxford/*Nature Picture Library*
130tl *USFWS*
130b Lynda Richardson/*Corbis*
134 E.Widder/*Harbor Branch Oceanographic Institution*
136 Captain Albert E. Theberge/*NOAA*
139 Florida Keys National Marine Sanctuary/*NOAA*
140 Paul Chesley/Stone/ *Getty Images*
141 Florida Keys National Marine Sanctuary/*NOAA*
143 Anders Ryman/*Corbis*
144 William Boyce/*Corbis*
145tr Natalie Fobes/*Corbis*
146 Dave G. Houser/*Corbis*
147 Ecoscene/*Corbis*
148 Ric Ergenbright/*Corbis*
149 Yann Arthus-Bertrand/ *Corbis*
152 Charles D. Winters/ *Science Photo Library*
156 Jim Richardson/*Corbis*
159 Yann Arthus-Bertrand/ *Corbis*
160 Space Imaging/*NASA*
162 Martin Harvey/*Still Pictures*
163 Jurgen Freund/*Nature Picture Library*
164–5b Matthew Oldfield, Scubazoo/*Science Photo Library*
164tl *NOAA*
165t Anthony Johnson/ The Image Bank/ *Getty Images*
174–5 Charles E. Rotkin/*Corbis*
178–9 Greg Smith/*Corbis Saba*
182 Joel Simon/Stone/*Getty Images*
183 Doug Allan/*Nature Picture Library*
190tl *ENVISAT ESA*
190–1b Tibor Bognar/*Corbis*
191t Jeffrey L. Rotman/*Corbis*
195 ISSEarthkAM/*NASA*
198 Royalty free/*Corbis*
199 Landsat 7/*NASA*
200 JSC/*NASA*
202–3 Stephen Frink/*Corbis*
207 Christine Osborne/*Corbis*
210 JSC/*NASA*
217 Michael Van Woert/*NOAA NESDIS, ORA*
218 OAR/National Undersea Research Program (NURP), National Marine Mammal Lab/*NOAA*
220 Galen Rowell/*Corbis*